BASIC electrotechnology

Butterworths BASIC Series includes the following titles:

BASIC aerodynamics
BASIC artificial intelligence
BASIC business analysis and operations research
BASIC business systems simulation
BASIC differential equations
BASIC economics
BASIC electrotechnology
BASIC fluid mechanics
BASIC hydraulics
BASIC hydrodynamics
BASIC hydrology
BASIC interactive graphics
BASIC investment appraisal
BASIC materials studies
BASIC matrix methods
BASIC mechanical vibrations
BASIC molecular spectroscopy
BASIC numerical mathematics
BASIC operational amplifiers
BASIC soil mechanics
BASIC statistics
BASIC stress analysis
BASIC theory of structures
BASIC thermodynamics and heat transfer

BASIC electrotechnology

R A Ashen, PhD, DIC, CEng, MIEE

Senior Lecturer, Power Electronics and Electric Drives Group,
Royal Military College of Science (Cranfield), Shrivenham, England

Butterworths
London   Boston   Durban   Singapore   Sydney   Toronto   Wellington

TK
153
.A76
1987

First published 1987

© **Butterworth & Co. (Publishers) Ltd, 1987**

British Library Cataloguing in Publication Data
Ashen, R.A.
  BASIC electrotechnology.—(Butterworths
  BASIC series).
  1. Electronics—Data processing   2. BASIC
  (Computer program language)
  I. Title
  621.381′028′55133   TK7835

  ISBN 0–408–01251–X

Library of Congress Cataloging-in-Publication Data
Ashen, R.A. (Richard Anthony)
  BASIC electrotechnology.
  (Butterworths BASIC series)
  Includes bibliographies and index.
  1. Electric engineering—Data processing.
  2. BASIC (Computer program language)   I. Title.
  II. Series.
  TK153.A76   1987   621.3′028′55133   87–13838
  ISBN 0–408–01251–X

Photoset by August Filmsetting, Haydock, St. Helens, Lancs.
Printed and bound in England by Page Bros Ltd, Norwich, Norfolk

Preface

This book is one of an expanding series devoted to the application of the computer language BASIC to engineering and mathematical subjects. The purpose of the book is twofold: to introduce students to the use of BASIC in an engineering context, and to show how BASIC may be applied to problems in electrotechnology.

The book is primarily intended for first-year common engineering and second-year electrical/electronic engineering undergraduates. However, the early chapters should not be beyond the ability of good mathematics and science students in schools.

BASIC was justifiably chosen as the programming language for this series of books for a variety of reasons. It is easy to learn and probably provides the quickest means of developing small programs to solve simple mathematical problems. In this respect, it is invaluable for engineering students keen to obtain solutions, who have limited interest in computing for its own sake. The major difficulties encountered should be engineering ones and not programming ones. Hence, programs may be quickly written and modified, and the student's confidence in his computing ability should be rapidly established. Since the series was started, the use of microcomputers in both school and home has increased dramatically. As BASIC is available on virtually all these machines, students should experience little difficulty in gaining access to suitable facilities to develop BASIC programs.

Despite its simplicity, the power of the language should not be underestimated. Relatively complex problems may now be readily solved at home or school with the help of microcomputers which, a few years ago, would have required the use of physically much larger minicomputers or 'mainframe' machines.

Since it was first conceived, BASIC has been enhanced in various ways, and many different versions of the language are now available. All programs in this book are written in 'minimal' BASIC, which corresponds closely with the original language formulation. Minimal BASIC was chosen for two reasons. Firstly, all versions of the language are compatible with minimal BASIC, which means that the programs should run on any computer supporting BASIC without modification. Secondly, it was intended to demonstrate that use of the most elementary version of the language imposes no significant restriction on the class or difficulty of mathematical problem that may be tackled. Some of the more complex programs could be shortened and improved by use of improved features available in more advanced versions of the language. Indeed, students may wish to modify programs where appropriate to make use of enhanced BASIC statements available to them. At no time, however, was the use of minimal BASIC found to preclude the tackling of any problem.

At this point, it is appropriate to mention the question of structured programming. BASIC is often criticised for its lack of structure, and for this reason it is frequently condemned as a 'bad' language. A structured program is essentially one which has a logical path through it, which passes through a succession of sections or 'blocks'. Writing programs in this logical block form makes them both easy to follow, and (relatively) easy to test for errors. Whilst it is true to say that BASIC does not encourage a structured approach, it does not preclude one, and a programmer capable of developing a structured program in a structured language should be able to employ the same essential approach when using BASIC. The importance of good programming style obviously increases as longer programs are developed, but it is beneficial to employ a disciplined approach to even the shortest program. This includes the use of appropriate and individual variable names which are easily remembered, and some form of program description or documentation.

Electrotechnology is a large subject, and it is not possible to cover all aspects of it in a book of this size. *BASIC electrotechnology* is not intended as a replacement for any of the existing textbooks on the subject, but is proposed rather to supplement them. Topics which best lend themselves to the BASIC approach have been concentrated on at the expense of others. In particular, with the exception of two examples in Chapter 6, the important subject of electrical measuring instruments has been neglected.

Chapter 1 forms an introduction to BASIC and describes the language in sufficient detail to follow the programs appearing later in the text. The subsequent chapters each have the same format. Each provides the essential theory for its own particular topic. A number of worked examples then follow, consisting of posed problems and the BASIC programs to solve them. Each program has a reasonably full set of notes describing its operation, and sample output. The significance of the results obtained are emphasised as much as possible. In some cases, worked examples have been used to introduce some new aspect of the theory. Occasionally, a question has been posed with a worked example to encourage the student to consider the implications of the solution obtained.

Many of the programs produce output suitable for graphical display. Minimal BASIC contains no specific graphics facilities, and so graphical output, where provided, has been produced manually. Students using computers with good graphics facilities may wish to improve the output sections of these programs. In common with many BASIC programs, and also hand calculators, output data is frequently displayed in this book to many decimal places of accuracy. This is mathematically unsatisfactory, but not over-important, provided that the data is interpreted with care. In particular, students should remember that output data can be no more accurate than the input data from which it was derived.

Following the worked examples, a number of problems are supplied for the student to tackle. These require the extension or application of existing programs or the development of new ones.

The use of complex numbers in circuit analysis is covered in Chapter 2. Chapter 3 describes linear circuit analysis with d.c. and sinusoidal a.c. supplies. Elementary magnetic circuit theory is presented in Chapter 4, leading on to the theory and performance of two-winding transformers from an equivalent circuit approach in Chapter 5. The final chapter forms an introduction to electromechanical energy conversion.

It is a pleasure to acknowledge colleagues who have assisted, both wittingly and unwittingly, with the writing of this book. In particular, thanks are due to the technical editor Mr P. D. Smith for sage and valid advice on various aspects of its content and production. For encouragement and advice, I am grateful to Dr R. E. Colyer, head of the Power Electronics and Electric Drives Group at RMCS. My colleagues in the group, Mr W. B. Lingard and Dr P. R. McLellan, also provided helpful advice on aspects of circuit analysis.

Finally, thanks are due to Miss S. Boss and Miss K. Farrow who, between them, word-processed the majority of the manuscript.

R.A.A.

In memory of my father

Contents

Introduction to BASIC

1.1 The BASIC approach

All the programs in this book are written in the programming language BASIC. BASIC (Beginner's All-purpose Symbolic Instruction Code) as first conceived differs from the majority of other high-level computer languages encountered in science and engineering, in being *interpreted*, rather than *compiled*. High-level scientific languages use statements resembling algebraic equations to make them easy to learn and use. However, these statements must be converted into a form intelligible to the computer before they can be implemented. Compiled languages, such as ALGOL, FORTRAN or PASCAL, perform this process once to produce a machine-code version which is invoked each time the program is run, or *executed*. For an interpreted language such as BASIC, however, the statements are decoded each time they are used during the run. Evidently, time is wasted in repeated interpretation of the same statement whenever a program is repeating a series of instructions, or 'looping'. Hence, BASIC is ideal for beginners writing short programs, since the effects of changes may be seen quickly, and student interest maintained. For longer programs, and where running speed is important, compiled languages are more efficient.

Many enhanced versions of BASIC have been developed since the language first appeared in 1964. Some computers now use compiled versions which, whilst improving program efficiency, lose the benefits of simplicity and immediacy which are important considerations for novice programmers.

The main features of the minimal version of BASIC employed in this book are described below. This is not a comprehensive specification of the language, being merely intended to enable the reader to understand the various types of statement he will encounter in the programs supplied. For a detailed description of minimal BASIC and other enhanced versions of the language, the interested reader is referred to any of the books listed in the bibliography at the end of this chapter.

1.2 The elements of BASIC

1.2.1 Mathematical expressions

The main purpose of the programs developed in the book is to solve equations developed for the various equipment and devices encountered in electrotechnology. This involves the use of variables and constants, which are all assigned names. No distinction is drawn between real and integer values, and an exponential form may be used to represent very large or very small numbers (e.g. 8.85E–12 represents 8.85×10^{-12}). Numeric variables are represented by names consisting of a letter or a letter followed by a digit (e.g. A and F1). Functions are supplied as standard to perform many common operations. Typical of these is SQR(X) to evaluate the square root of X, and SIN(X) to calculate the sine of an angle X in radians. The argument in brackets (X) may be a number, a variable, or a mathematical expression. A full list of the minimal BASIC supplied functions is given at the end of the chapter.

Mathematical equations also contain operators such as plus, minus, etc. These operators have an order in which they are evaluated, this being

exponentiation ($^$) (first)
multiply ($*$) and divide ($/$)
add ($+$) and subtract ($-$) (last).

For operators of equal precedence, the computer works from left to right. As in algebra, brackets may be used to override the precedence. Hence $\dfrac{(A+B)}{3C}$ becomes $(A+B)/(3*C)$ or $(A+B)/3/C$.

1.2.2 Program structure and assignment

A BASIC program is a sequence of statements which define a procedure for the computer to follow. The statements occupy one line each, the lines having unique numbers defining the order in which the statements are executed. Generally, programs in this book have statement numbers in the ascending order 10, 20, 30 etc. As the program proceeds, values are assigned to each of the variables. Some values may be specified by input data or may be preset by the program. Others are generated in the program by use of assignment statements. These have the form

line number [LET] variable = mathematical expression

where the square brackets indicate that the word LET is optional, and it is therefore usually omitted (it has been in this book). As an example, the modulus **Z** of a complex number $x+jy$ is given by $\sqrt{(x^2+y^2)}$. A BASIC assignment statement to perform this calculation is

50 Z = SQR(X^2 + Y^2)

It is important to realise that an assignment statement is not itself an equation. Rather, it is an instruction to the computer to assign the variable on the left-hand side of the equation the numeric value of the expression on the right-hand side. Hence, it is permissible to have the assignment statement

60 X = X + 1

which increases by 1 the value of X.

Variables may be assigned constant values, which remain unchanged throughout the program. One way is by use of an assignment statement such as

30 P1 = 3.14159265

An alternative, which is particularly useful if many constants are involved, is to use DATA and READ statements. These take the form

line number DATA number 1 [,number 2, . . .]
line number READ variable 1 [, variable 2, . . .]

For example, the following statements assign the values $-273, 0$, and 100 to variables T0, T1, and T2 respectively

100 DATA −273, 0, 100
110 READ T0, T1, T2

If the data list is shorter than the list of READ variables, then an error is normally indicated by the computer. DATA statements may be placed anywhere in the program, but it is good practice to place them at the beginning, together with their associated read statements, where they are easily found for amendment if necessary. It is sometimes required to give the same set of data values to more than one set of variables. This may be achieved simply by use of the RESTORE statement, which resets the data pointer to the start of the data list. Hence, if the values 1, 2, and 3 are to be assigned to both A, B, and C respectively and X, Y, and Z respectively, this could be achieved by the following statements

20 DATA 1, 2, 3
30 READ A, B, C
40 RESTORE
50 READ X, Y, Z

Non-numeric data (e.g. words) can be handled by string variables. A string is a character or series of characters within quotes (e.g. 'I', 'CURRENT', 'VOLTAGE'). A string variable is a letter followed by a $ sign. Another valid DATA/READ combination is therefore

```
60 DATA "RED", "ORANGE", "YELLOW"
70 READ C$, D$, E$
```

1.2.3 Input

For interactive or 'conversational' programs, the user specifies variables by inputting data in response to prompts from the computer as the program is running. The INPUT statement has the form

line number INPUT variable 1 [, variable 2, . . .]

e.g.

```
20 INPUT A, B, C
```

When the program is run, the program prints ? as it reaches this statement and waits for the user to type values for the variables, e.g.

?1,2,3

which assigns the variables the same values as before.

1.2.4 Output

Output of data and the results of calculations, etc. is implemented by using a statement of the form

Line number PRINT list

where the list may contain variables or expressions, e.g.

```
200 PRINT A, B, C, A*B/C
```

text enclosed in quotes, e.g.

```
50 PRINT "Current (A)      Force (Nm)"
```

or mixed text and variables, e.g.

```
500 PRINT "Resistance = "; R; "(Ohm)"
```

Variable values begin with a space if positive, or a minus sign if negative. The items in the list are separated by commas or semi-colons. Commas give tabulation in columns. The column width varies from computer to computer, but is typically about 15 spaces. A semi-colon suppresses this spacing, and a semi-colon at the end of a list suppresses the line feed. A TAB may be used to force printing to start in a particular column. For example, if the word 'Resistance' in the previous example is to begin in column 10, then this could be achieved by the following PRINT statement

500 PRINT TAB(10); "Resistance = "; R; "Ohm"

A PRINT statement with no list prints a blank line. Blank lines are a useful means of improving output layout and they have been widely used in the programs in this book.

It is important in general to use PRINT statements with both 'run-time' input and READ/DATA statements, so that the program user can maintain a record of the data used.

1.2.5 Conditional statements

It is often necessary to enable a program to take some action if, and only if, some condition is fulfilled. This is done with a statement of the form

line number IF expression 1 conditional operator expression 2 THEN [GOTO] line number

where the possible conditional operators are

= equals
< > not equal to
< less than
< = less than or equal to
> greater than
> = greater than or equal to

For example, a program could contain the following statements if it is to stop when a zero value for A is input

20 INPUT A
30 IF A < > 0 THEN 50
40 STOP
50 . . .

The statement

line number STOP

stops execution of the program. It is the only correct way of terminating program operation. The statement

line number END

may be used as the last line of the program, if desired. The END statement is optional, but shows clearly that no more statements are expected.

1.2.6 Loops

There are several means by which a program can repeat some of its procedure. The simplest means is by use of the GOTO statement which has the form

line number GOTO line number

However, the GOTO statement in this form should be used sparingly, since excessive use soon makes programs difficult to follow. A better way of performing loops is with a starting statement of the form

line number FOR variable = expression 1 TO expression 2 [STEP] expression 3

where the step is assumed to be unity if omitted. The finish of the loop is signified by the statement

line number NEXT variable

where the same variable is used in both FOR and NEXT statements. Its value must not be changed in the intervening lines. Note that the step may be positive or negative, and that the loop is terminated when the variable exceeds expression 2, not when it equals it. It is permissible to place loops inside each other in 'nested' form, but they must not interleave each other. FOR/NEXT loops should be entered at the FOR and left at the NEXT; it is permissible to branch from an inner loop into an outer loop, but this is generally bad practice unless branching to an error statement. Indenting of FOR/NEXT loops makes programs easier to follow, especially if loops are nested to any depth. The following, then, is correct BASIC

```
100 FOR J = 1 TO 4
110    FOR K = 10 TO 2 STEP − 2
       ......
       ......
180    NEXT K
       ......
       ......
250    FOR K = 0.1 TO 3.3 STEP 0.1
       ......
       ......
290       FOR L = 10 TO 100 STEP 10
          ......
          ......
```

350   NEXT L
360   NEXT K
. . . . . .
. . . . . .
500   NEXT J

Note, however, that if the statement order of NEXT L and NEXT K at lines 350 and 360 had been reversed, then the corresponding FOR/NEXT loops would be interleaved, which would be incorrect.

1.2.7 Subscripted variables

When dealing with many similar quantities, it is convenient to be able to refer to them collectively with the same variable name. This is achieved by the use of subscripted variables. Hence, a number of electric currents can be given the collective name I, the individual currents being $I(0)$, $I(1)$, $I(2)$ etc. Such a collection of similar quantities is called an array. The use of arrays facilitates mathematical operations with these multiple quantities. For example, the currents above might be calculated from corresponding arrays of voltages, V, and resistances R by BASIC statements of the form

250 FOR N = 1 TO 10
260 I(N) = V(N)/R(N)
270 NEXT N

In order to allocate storage space for arrays, the computer must be told how many subscripted variables each contains. This is done using the DIMENSION statement, which has the form

line number DIM variable 1 (integer 1) [, variable 2 (integer 2), . . .]

Each subscripted variable must have an associated DIM statement which must appear before the variable is first used. These variables may have more than one dimension. For example, the statement

20 DIM I1 (5, 10)

could be used to reserve space for currents whose values are to be specified at successive instants in time. Note that, in general, array subscripts start from zero (not one), and that the number specified in the DIM statement is the maximum subscript allowed (not the number of elements in the array). Hence, the above array could hold six currents and 11 time instants.

In some systems an OPTION BASE statement allows the minimum subscript to be set to 1. In such a system, the appearance of the statement

10 OPTION BASE 1

before the above DIM statement would restrict the array to holding five currents and 10 time instants, starting at I1 (1,1) and finishing at I1 (5,10).

1.2.8 Subroutines

Sometimes a sequence of statements needs to be accessed more than once in the same program. Instead of merely repeating these statements, it is better to put them in a subroutine. The program then contains statements of the form

line number GOSUB line number

When the program reaches this statement, it branches (i.e. transfers control) to the second line number. The sequence of statements starting with this second line number ends with a statement

line number RETURN

and the program returns control to the statement immediately after the GOSUB call.

Subroutines may be placed anywhere in the program, but it is usually convenient to position them at the end, separated from the main program by a STOP statement.

Subroutines may be nested in a similar manner to FOR/NEXT loops. In this situation, RETURN always returns control to the statement immediately following the GOSUB call to its own routine.

The use of subroutines is to be encouraged, particularly in large programs, since it breaks programs down into individual blocks, or modules. It is easier to test and maintain a program so written, and subroutines written for one program may often be used in another. (e.g. see example program 5.3 (TRANSCT)).

1.2.9 User-defined functions

In addition to those supplied with the system, the user can define his own functions with the DEF FN statement which has the form

line number DEF FN-letter [(variable)] = expression

and typical forms are

10 DEF FNE = 2.71828
20 DEF FNC(A) = 1/TAN(A)
30 DEF FNM(T) = 2*(COS(T))^2 − 1

The form containing the variable is the more useful. The function is invoked in a similar manner to the supplied function. Hence with the definitions above, the statement

50 Y = FNC(0.5)

would assign the variable Y the value of the cotangent of 0.5 radians. Note that the variable in the DEF statement has no meaning outside the function. Hence the variables A and T could be used elsewhere in the program without any resulting difficulty.

It is usually necessary for the DEF FN statement to appear before the function is used. (This rule has, in fact, been broken with the one user-defined function employed in this book in Example 6.2 (WMETER)).

1.2.10 Other statements

(1) Explanatory remarks or comments which are not to be output can be inserted into a program using

line number REM comment

Any statement beginning with the word REM is ignored by the computer.

It is good practice to include REM statements whilst developing programs. This benefits both the program writer, and anyone else who subsequently may wish to understand the program.

(2) Multiple branching may be achieved with a statement of the form

line number ON expression GOTO 1st line number [,2nd line number,...]

When a program reaches this statement, it branches to the first line number if the integer value of the expression is 1, to the second number if the value is 2, and so on. An error message may be printed, or the program may proceed to the next statement, if the expression gives a value less than 1 or greater than the number of referenced line numbers. (The ON GOTO statement has not been used in this book.)

1.3 Checking programs

Errors of various types may occur during program development, and the process of producing a 'correct' program is termed 'debugging'. If a syntax error is present, i.e. an incorrect BASIC statement has been input, then the program will not run and a relevant message is normally displayed by the computer. These errors are usually fairly easy to eradicate.

More difficult to detect are logic errors, where all the rules of

BASIC have been obeyed, but the program gives incorrect results. Logic errors may be introduced by incorrect program formulation, or mis-typed mathematical operators or variable names. These errors should be checked for by running the program with data which gives a known solution, or by checking against a hand calculation. In complex programs, many test results may have to be checked to ensure program reliability, since there may be many alternative paths. Having detected the presence of a logic error, one technique to locate it is to insert temporary PRINT statements to print values of variables at critical points throughout the program. These PRINT statements may be removed once the error has been located and corrected.

Good program structure is of assistance in program debugging. The advantages of subroutines in this respect have already been mentioned. These assist in breaking the program down into subsections, each of which may be tested separately for correct operation.

1.4 Summary of minimal BASIC statements

Assignment

LET	Computes and assigns value
DIM	Allocates space for subscripted variables (arrays)
READ	Reads data from DATA statements
DATA	Storage area for data
RESTORE	Restores data pointer to start of DATA

Input

INPUT	Reads data from 'run-time' keyboard input

Output

PRINT	Prints output list

Program control

IF ... THEN	Conditional branching
GOTO	Unconditional branching
FOR ... TO ... STEP	Opens loop
NEXT	Closes loop
GOSUB	Transfers control to subroutine
RETURN	Return from subroutine
ON ... GOTO	Multiple branching
STOP	Stops program execution
END	Last line of program (optional)

Comment

REM   Comment in program

Functions

ABS   Absolute value (modulus)
ATN   Arctangent (gives angle in radians)
COS   Cosine (angle in radians)
EXP   Exponential
INT   Largest integer not exceeding value
LOG   Natural logarithm
SGN   Sign of value
                           (-1 if value <0
                             0 if value $=0$
                             $+1$ if value >0)
SIN   Sine (angle in radians)
SQR   Square root
TAN   Tangent (angle in radians)
DEF FN   User-defined function

(*Note* The computer used to develop the programs in this book contained a natural logarithm function named LN, rather than LOG, which is the standard name as specified above.)

1.5 Bibliography

MONRO D. M., *Interactive Computing with BASIC*, Arnold, London (1974)
KOFFMANN E. B. and FRIEDMAN F. L., *Problem Solving and Structured Programming in BASIC*, Addison-Wesley, London (1979)
HOLMES B. J., *BASIC Programming*, D.P. Publications, Eastleigh (1982)

Use of complex numbers in a.c. circuit analysis

ESSENTIAL THEORY

2.1 Introduction

A.C. circuits are frequently fed from supplies which are, to a good approximation, sinusoidal. If all circuit components are mathematically linear, then all currents and voltages appearing in the circuit will also vary sinusoidally. Solutions may be obtained for all these time-varying quantities, using various circuit theorems, in terms of sinusoidal functions, but the mathematical manipulation may be greatly simplified by the use of complex notation. The introduction of complex notation has the added advantage that a.c. circuits may then be handled directly by the various techniques developed for d.c. circuits.

2.2 Complex numbers

Complex numbers are frequently first encountered in the solution of quadratic equations, where the lack of meaning for the square root of a negative number in the real number system introduces a problem. This problem is overcome by defining these square roots as imaginary numbers. A general combination of a real and an imaginary number is then defined as a complex number. The most basic representation of a complex number is hence

$$\mathbf{Z} = x + jy \tag{2.1}$$

where x and y are real numbers and $j = \sqrt{-1}$. (j is used in this text, rather than the symbol i favoured by mathematicians, to prevent confusion with the symbol for current). Complex numbers are indicated here by the use of heavy type. x is termed the *real part* of the complex number, and y the *imaginary part*. Complex numbers may be conveniently represented graphically on an Argand diagram, as shown in Figure 2.1. Pure real numbers occur on the horizontal, or real, axis, and pure imaginary numbers on the vertical, or imaginary, axis.

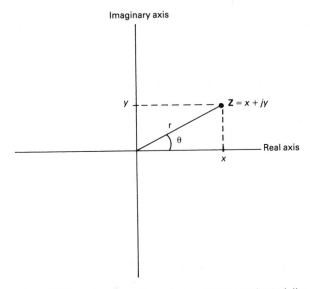

Figure 2.1 Representation of complex numbers on an Argand diagram

It is clear from the diagram that an alternative representation of the complex number is possible, based on polar co-ordinates. Since if

$$x = r \cos \theta \tag{2.2}$$

and

$$y = r \sin \theta \tag{2.3}$$

then

$$\mathbf{Z} = r (\cos \theta + j \sin \theta) \tag{2.4}$$

This polar form is conventionally written as

$$\mathbf{Z} = r \angle \theta \tag{2.5}$$

where r is termed the *modulus* and θ the angle or *argument* of the complex number.

A third representation of complex numbers, of particular use for multiplication and division, is also possible. It may be readily demonstrated by power series expansion that

$$\cos \theta + j \sin \theta = e^{j\theta} \tag{2.6}$$

This identity then results in the exponential form of the complex number

$$\mathbf{Z} = r \, e^{j\theta} \tag{2.7}$$

(a)

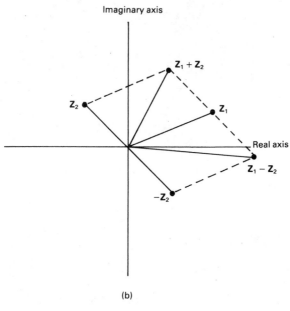

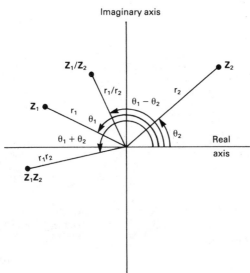

Figure 2.2(a) complex addition and subtraction   (b) complex multiplication and division

For circuit-analysis purposes, operations of addition, subtraction, multiplication and division are needed using complex numbers. Addition and subtraction are readily carried out with complex numbers expressed in basic form. If $\mathbf{Z}_1 = a + jb$, and $\mathbf{Z}_2 = c + jd$ then

$$\mathbf{Z}_1 + \mathbf{Z}_2 = (a + c) + j(b + d) \tag{2.8}$$

with the real and imaginary parts added (or subtracted) separately. This process corresponds to vector addition or subtraction using the 'parallelogram of forces', and may be carried out on the Argand diagram, as shown in Figure 2.2(a).

Multiplication and division may be performed directly with complex numbers in polar or exponential form. If $\mathbf{Z}_1 = r_1 \angle \theta_1$ and $\mathbf{Z}_2 = r_2 \angle \theta_2$ then

$$\mathbf{Z}_1 \mathbf{Z}_2 = r_1 r_2 \angle (\theta_1 + \theta_2) \tag{2.9}$$

and

$$\mathbf{Z}_1 / \mathbf{Z}_2 = (r_1/r_2) \angle (\theta_1 - \theta_2) \tag{2.10}$$

and so the rule is multiply (or divide) the modulii, and add (or subtract) the angles. These operations are illustrated in Figure 2.2(b). Since $j = 1 \angle \pi/2$ in polar form, then $j^2 = 1 \angle \pi$ from the above definition. $1 \angle \pi = -1$, as can be seen from the Argand diagram, and so the definitions of j and complex multiplication are consistent.

The complex conjugate $\mathbf{Z}^*$ of the complex number $\mathbf{Z} = x + jy$ is defined as

$$\mathbf{Z}^* = x - jy \text{ (or } r \angle -\theta \text{ or } re^{-j\theta}) \tag{2.11}$$

Hence it is the reflection of $\mathbf{Z}$ in the real axis of the Argand diagram. It follows from the multiplication definition that

$$\mathbf{Z}\mathbf{Z}^* = r^2 \tag{2.12}$$

a real number.

The complex conjugate is used in calculations of power in a.c. circuits.

2.3 Representation of sine waves by complex notation

Figure 2.3(a) represents two quantities varying sinusoidally in time at the same frequency f. The time period T to complete one cycle is evidently $1/f$, where f is measured in cycles/s which has been given the unit name Hertz (Hz). It is more convenient mathematically to work in terms of angular frequency ω, measured in rad/s. From the definition of the radian, it follows directly that

$$\omega = 2\pi f \tag{2.13}$$

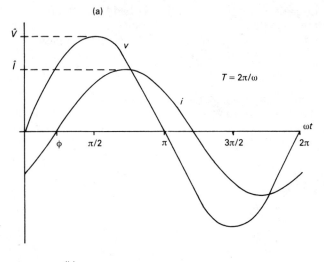

(a)

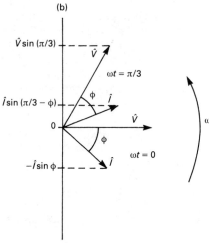

(b)

Figure2.3(a) phase-displaced sine waves at same frequency   (b) complexor representation

The equations of the two sine waves in Figure 2.3(a) are

$$v = \hat{V} \sin \omega t \tag{2.14}$$

and

$$i = \hat{I} \sin (\omega t - \varphi) \tag{2.15}$$

Since v reaches its peak value before i in time, i is said to *lag* v, or conversely v is said to *lead* i. The *phase angle* is a measure of the phase displacement between the sine waves, which are said to be *out of phase*. For the particular case that $\varphi = 0$, the waveforms are coincident in time, or *in phase*.

v and i may be represented by rotating lines of fixed length, called *complexors*, as indicated in Figure 2.3(b). With line lengths proportional to peak values, and anticlockwise rotation at angular frequency ω, projections of the complexors on to the vertical axis give instantaneous values of v and i. The situation at two time instants is indicated in the diagram. Since v and i have the same frequency, the complexors maintain a constant angular separation, which is the phase angle φ. Hence the sine waves may be conveniently and simply represented by the complexors frozen at some instant in time. One complexor is considered as the reference or datum from which others are derived, and this is drawn, by convention, along the positive x axis of the diagram. Complexors associated with time-varying sinusoidal quantities are termed *phasors*, and the diagram resulting from this process is termed a *phasor diagram*.

The phasor diagram, taking v as reference, is shown in Figure 2.4. Comparison with Figures 2.1 and 2.2 indicate the similarity with the Argand diagram, and hence how phasors may be conveniently represented mathematically by complex numbers. The polar forms for $\mathbf{V}$ and $\mathbf{I}$ are

$$\mathbf{V} = \hat{V} \angle 0 \qquad (2.16)$$
$$\mathbf{I} = \hat{I} \angle -\varphi \qquad (2.17)$$

From the electrical viewpoint, the effective value of a sine wave is its *root mean square* (r.m.s.) value, since it may be readily demonstrated that the r.m.s. value of any alternating quantity is the corresponding steady direct quantity associated with the same power. Hence, phasors are normally scaled in terms of r.m.s. rather than peak values. The ratio peak/r.m.s. value for a sine wave is $\sqrt{2}$.

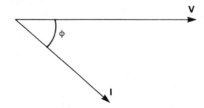

Figure 2.4 Phasor diagram for v and i

2.4 Resistance

If a potential difference exists between the ends of a piece of metal, free electrons present in the material move from the lower to the higher potential. This electron movement constitutes an electric current which, by convention, is considered to flow in the other direction, i.e. from high to low potential. If all other conditions are held constant, it is found that the current flowing is proportional to the potential difference applied. This is *Ohm's law*, and the ratio potential difference in volts (V) to current in Ampères (A) is defined as the *resistance* R of the metal in Ohms (Ω).

Electrons are particles of negative electric charge, and current is related to the rate of charge movement. The unit of electric charge is the Coulomb (C), and the Ampère corresponds to a rate of charge movement of 1 C/s.

For most materials that conduct electricity, it is found that electrical resistance is proportional to length of conduction path l, and inversely proportional to the cross-sectional area of the path A. Hence

$$R = \rho l / A \tag{2.18}$$

The constant of proportionality, ρ, is the *resistivity* of the material, with unit Ωm.

Alternatively, the *conductance* G of the material may be considered, where $G = 1/R$ and

$$G = \sigma A / l \tag{2.19}$$

The unit of conductance is the Siemen (S) and σ is the *conductivity* of the material with unit S/m.

Resistivity values at room temperature vary from 1.62×10^{-8} Ωm for silver through the range 10^{-4}–10^{2} Ωm for semiconductor materials and up to 10^{17} Ωm for the best electrical insulators.

Resistivity varies considerably with temperature. In conductors, resistivity rises with increased temperature because increased thermal agitation of the free conduction electrons causes more collisions and reduces their mobility. Over small temperature ranges, the variation is reasonably linear. In contrast, semiconductors, insulators and liquid electrolytes typically show a decrease in resistivity with increased temperature due to increased production of charge carriers and increased diffusion rates.

2.5 Inductance and inductive reactance

Electric current is movement of electric charge, and there is always a magnetic field associated with an electric current. The magnetic field is represented by lines of magnetic flux which form closed paths around the closed current-carrying electric circuits. Arrows drawn on the flux lines indicate the direction in which a (fictitious) unit north magnetic pole would move. With this convention, current and flux are linked in the sense of a right-hand corkscrew or screw thread. Magnetic flux is also associated with permanent magnets. Magnetic flux is given the symbol Φ and its unit is the Weber (Wb).

Some commonly encountered flux/current patterns are illustrated in Figure 2.5. It is often convenient to show sections through current or flux paths when sketching circuits. In these situations, the 'dot' and 'cross' convention is used to indicate direction, with the dot indicating direction 'out of the paper' as the point of an arrow, and the cross indicating direction 'into the paper' as the tail of an arrow.

Inductance is a measure of an electric circuit's ability to set up

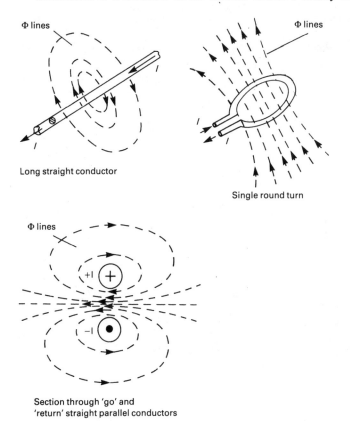

Long straight conductor

Single round turn

Section through 'go' and
'return' straight parallel conductors

Figure 2.5 Typical magnetic flux/electric current relationships

magnetic flux. For a fixed current, flux may be increased by winding wire into a coil (see Figure 2.6(a)) and, to a far greater extent, by including magnetic iron or steel in the magnetic circuit (see Figure 2.6(b)).

Hence, inductance is usually associated with coils, but it is important to realise that every electric circuit possesses some inductance, whether desirable or not, because some magnetic flux is always set up when a current flows.

Inductance, L, of a coil or circuit is given by the formula

$$L = N\Phi/I \tag{2.20}$$

where N is the number of turns linked by the flux. The unit of inductance is the Henry (H). The quantity $N\Phi$ is termed the *flux linkage* of the circuit or coil, and is given the symbol Ψ.

Although the expression for inductance is simple, it can prove difficult to calculate in cases where the linkage between flux and turns is not straightforward.

To evaluate the effect of inductance in an electric circuit, it is necessary to relate the current through it to the potential difference (voltage) between its ends. This relation follows from *Faraday's law of electromagnetic induction* which states that when the flux linking a

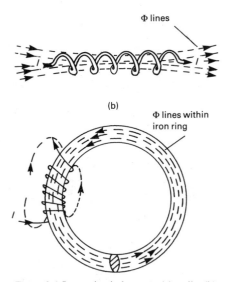

Figure 2.6 Increasing inductance (a) coil   (b) coil on iron ring

coil changes, an *electromotive force* (e.m.f.) E is induced, equal to the rate of change of flux linkage. The polarity of the induced e.m.f. is given by *Lenz' law* which states that the induced e.m.f. will tend to set up a current to oppose the change in flux linkage. Both laws are incorporated in the equation

$$E = -N\frac{d\Phi}{dt} = -\frac{d\Psi}{dt}$$ (2.21)

But, if inductance is constant (this may not be the case if magnetic material is present), it follows from equation (2.20) that

$$E = -L\frac{dI}{dt}$$ (2.22)

This is the required voltage/current relationship for circuit purposes.

It should be evident that inductance only influences circuit performance when current is changing, and that a pure inductance cannot be produced, since any circuit or coil must contain some additional resistance.

With sinusoidal supplies, the current waveform is given by an expression of the general form $\sqrt{2}I\sin(\omega t - \varphi)$, where I is now the r.m.s. value. Hence from Equation (2.22), this current would produce an induced voltage of $-\sqrt{2}\omega LI \cos(\omega t - \varphi)$ across a pure inductance. The applied voltage v is equal and opposite to this, as shown in Figure 2.7(a). Note that the current lags the applied voltage by a quarter of the a.c. cycle, or $\pi/2$ rad. For circuit analysis, it is necessary to relate the current and voltage mathematically. Clearly, the r.m.s. value of the applied voltage is ωLI, and so the ratio of r.m.s. voltage to current is ωL. However, the phase relationship of $\pi/2$ rad must also be represented in the ratio. This is readily achieved using phasor notation and complex arithmetic. The appropriate phasor diagram, scaled in r.m.s. values, is shown in Figure 2.7(b).

From the phasor diagram

$$\frac{V}{I} = \frac{\omega LI \angle (\pi/2 - \varphi)}{I \angle -\varphi} = \omega L \angle \pi/2 = j\omega L$$ (2.23)

The quantity ωL is called the *inductive reactance*, and is given the symbol X_L. It is the a.c. equivalent of 'resistance' for an inductance and equation (2.23) corresponds to Ohm's law for an inductance. The presence of j in the equation represents the $\pi/2$ phase displacement between V and I. Note that the complex arithmetic depends on the *relative* phase displacement between the phasors, and so it is unaffected by which, if either, is chosen as reference.

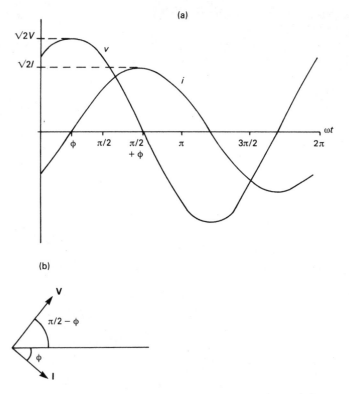

Figure 2.7 Voltage-current relationship for inductance (a) time variation
(b) phasor diagram

2.6 Capacitance and capacitive reactance

A capacitor is a device which stores electric charge, consisting, in its
simplest form, of a pair of metal plates, separated by a slab of insulating material, called a *dielectric* (see Figure 2.8(a)). If the capacitor is
connected to a d.c. source, such as a battery, via a closed switch,
electrons are attracted from one capacitor plate by the positive battery terminal, and flow through the battery and arrive on the other
capacitor plate, as indicated in Figure 2.8(b). The nett effect is to leave
the first plate positively charged, and to produce an equal and
opposite negative charge on the second plate. This process cannot
continue, however, since electrons cannot flow through the dielectric.
The charge separation results in an opposing potential difference
being developed across the capacitor, and electron movement ceases
when this potential difference reaches the battery e.m.f. In circuit

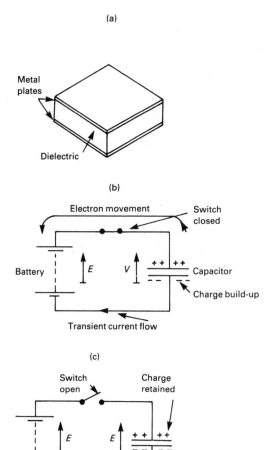

Figure 2.8(a) parallel-plate capacitor   (b) charging current from battery
(c) charge and potential retention when disconnected

terms, this process is represented by a momentary, or transient, current flow, driven by the battery.

The effect of the charged plates is to strain the molecules of the dielectric, such that the electrons in their bound atomic orbits are attracted towards the positively charged plate, without breaking free from their parent atoms. This effect is called *polarisation*, and it is the essential mechanism by which energy is stored in the capacitor.

If the switch is now opened, the charge remains on the capacitor plates because no current can flow around the incomplete circuit. Hence, as shown in Figure 2.8(c), the capacitor may be completely disconnected and, in theory, it would retain its charge, and associated potential difference, indefinitely. In practice, however, conduction paths between the plates result in charge leakage in limited time, and eventual complete discharge.

It is found that the charge held on either plate is directly proportional to the voltage, or potential difference between the plates. The ratio of charge to voltage is defined as the *capacitance, C*, of the device, and the unit assigned to capacitance is the Farad (F). Hence

$$Q = CV \tag{2.24}$$

for a capacitor. In practice, the farad is found to be a very large unit and so the microfarad ($\mu F = 10^{-6} F$) or picofarad ($\mu\mu F$ or pF $= 10^{-12} F$) are more commonly used.

For the parallel-plate capacitor considered, the capacitance is proportional to plate area, A, and inversely proportional to plate separation, d, and is governed by the equation

$$C = \frac{\varepsilon A}{d} \tag{2.25}$$

ε is a property of the dielectric termed its *permittivity*, with units F/m. The value of ε for a vacuum is a fundamental constant called the *permittivity of free space*. It is given the symbol ε_o and has a value of 8.85×10^{-12} F/m. This apparently odd value is in fact $1/(36\pi) \times 10^{-9}$ and arises from the way in which electrostatic units were initially defined.

It is convenient to relate permittivity of dielectrics to that of a vacuum by the *relative permittivity* ε_r where

$$\varepsilon = \varepsilon_r \varepsilon_o \tag{2.26}$$

ε_r is evidently a dimensionless quantity.

Values of relative permittivity attainable vary from approximately 1 for air, through 2 for paper, 2.8 for polycarbonate which is commonly used in modern capacitors, 6 for mica which was formerly commonly used, to values of several thousands for materials called *ferro-electrics* of which barium titanate is one example. Unfortunately, at present, stability problems preclude the widespread use of ferro-electrics in capacitors.

Since current is the rate of charge movement, Equation (2.24) may be differentiated with respect to time to give the following relationship between voltage and current for a capacitor.

$$I = C\frac{dV}{dt} \tag{2.27}$$

(a)

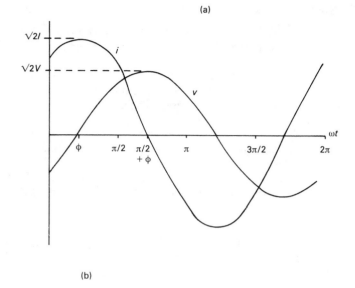

(b)

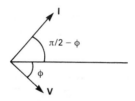

Figure 2.9 Voltage-current relationship for capacitance (a) time variation (b) phasor diagram

For sinusoidal supply conditions, a similar procedure may be carried out for capacitance as was performed for inductance, and this is illustrated in Figure 2.9. Here, voltage may be conveniently represented by the expression $\sqrt{2}V \sin(\omega t - \varphi)$, and the resulting current is $\sqrt{2}\omega CV \cos(\omega t - \varphi)$. These waveforms are shown in Figure 2.9(a). In this case, the current leads the voltage by $\pi/2$ rad, as opposed to the equivalent lag in the inductive case. The corresponding phasor diagram is shown in Figure 2.9(b). It should now be self-evident that

$$V = \frac{-j}{\omega C} I \qquad (2.28)$$

$1/(\omega C)$ is termed *capacitive reactance* and is given the symbol X_c. The presence of $-j$ in the equation indicates the $\pi/2$ rad lag of the voltage behind the current in the capacitor.

2.7 Impedance

The *impedance* of a circuit, or any individual circuit component, is defined as the ratio of potential difference between the ends of the circuit to current flowing through it.

Under sinusoidal a.c. conditions, impedance can be expressed using complex notation as has been previously developed. Hence it may be expressed as R, jX_L, and $-jX_c$ for resistance, inductance and capacitance respectively.

The benefits of complex notation become apparent when the impedance of circuits containing many elements is to be evaluated. Under sinusoidal supply conditions, elements may be combined in series and parallel, using complex notation, in a manner essentially identical to that employed in d.c. circuits. This approach is developed further in the following chapter.

2.8 Bibliography

MASON, J. C., *BASIC Numerical Mathematics*, Butterworths, London (1983)

WORKED EXAMPLES

Example 2.1: COMMUL: complex number multiplication

It has been explained in the text that a.c. circuit analysis requires basic mathematical operations to be performed with complex numbers. Write a program to perform multiplication on two complex numbers in the $x + jy$ (or Cartesian) form.

```
10 REM       COMMUL - COMPLEX Number Multiplication
20 PRINT
30 PRINT
40 PRINT "PRODUCT OF A PAIR OF COMPLEX NUMBERS"
50 PRINT "-------------------------------------"
60 PRINT
70 PRINT
80 PRINT "Input numbers in Cartesian (a + jb) form"
90 PRINT
100 PRINT
110 PRINT "REAL part of first complex number";
120 INPUT A
130 PRINT
140 PRINT "IMAGINARY part of first complex number";
150 INPUT B
160 PRINT
170 PRINT "REAL part of second complex number";
180 INPUT C
190 PRINT
200 PRINT "IMAGINARY part of second complex number";
```

```
210 INPUT D
220 R = A*C - B*D
230 I = B*C + A*D
240 PRINT
250 PRINT "REAL part of product is: ";R
260 PRINT "IMAG part of product is: ";I
270 PRINT
280 PRINT
290 PRINT "Do you wish to run the Program again (Y/N)";
300 INPUT Q$
310  IF Q$ = "Y" THEN GOTO 60
320 STOP
330 END
```

```
PRODUCT OF A PAIR OF COMPLEX NUMBERS
------------------------------------

Input numbers in Cartesian (a + jb) form

REAL part of first complex number?1

IMAGINARY part of first complex number?1

REAL part of second complex number?2

IMAGINARY part of second complex number?1

REAL part of product is: 1
IMAG part of product is: 3

Do you wish to run the Program again (Y/N)?N

STOP at line 320
```

Program notes

(1) Lines 110 to 210 cover the input of real and imaginary parts of the two complex numbers.

(2) Extra PRINT statements without variables or text are liberally distributed throughout the program. These introduce blank lines which are intended to improve the screen's readability.

(3) The real and imaginary parts of the product are calculated at lines 220 and 230 respectively. These are extracted from the equation:

$$R + jI = (A + jB) \times (C + jD) \tag{2.29}$$

(4) The string variable Q$ is used at lines 290 to 310 to test whether or not the program is to be run again.

If Y is input, the program jumps to line 60 and repeats with a request for more data. If N (or anything else) is input, the program terminates.

Example 2.2: CARPOL: cartesian to polar conversion

Complex number multiplication and division may be simplified if the numbers are in polar form. Write a program to convert a complex number from Cartesian to Polar form. The program must be able to correctly convert pure imaginary numbers and complex numbers with negative real parts. Angles are to be output in degrees.

```
10 REM      CARPOL - Cartesian to Polar Conversion
20 PRINT
30 PRINT
40 PRINT "CARTESIAN TO POLAR CONVERSION"
50 PRINT "-------------------------------"
60 PRINT
70 PRINT
80 PRINT "REAL part of complex number";
90 INPUT A
100 PRINT
110 PRINT "IMAGINARY part of complex number";
120 INPUT B
130   M = SQR(A^2 + B^2)
140 IF A = 0 THEN GOTO 230
150 IF A < 0 THEN GOTO 180
160 C = 57.296*ATN(B/A)
170 GOTO 280
180 IF B < 0 THEN GOTO 210
190 C = 57.296*(3.14159 + ATN(B/A))
200 GOTO 280
210 C = -57.296*(3.14159 - ATN(B/A))
220 GOTO 280
230 IF B > 0 THEN C = 90
240 GOTO 280
250 IF B = 0 THEN C = 0
260 GOTO 280
270 IF B < 0 THEN C = -90
280 PRINT
290 PRINT "Polar form is: ";M;" < ";C
300 PRINT
310 PRINT
320 PRINT "Another conversion (Y/N)";
330 INPUT Q$
340 IF Q$ = "Y" THEN GOTO 60
350 STOP
360 END
```

```
CARTESIAN TO POLAR CONVERSION
-------------------------------

REAL part of complex number?-1

IMAGINARY part of complex number?1

Polar form is: 1.41421356 < 135.000367

Another conversion (Y/N)?N

STOP at line 350
```

Program notes

(1) The program uses the supplied function ATN(X), which calculates the angle in radians whose tangent is X. The function returns values between $+\pi/2$ radians and $-\pi/2$ radians, with a positive value resulting from a positive X and a negative value from a negative X. ATN(B/A) is used to calculate the polar angle of the complex number $A + jB$. Hence complex numbers with positive real parts (in the first or fourth quadrants, as indicated in Figure 2.10) are within the standard

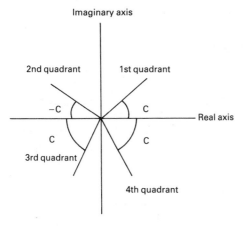

Figure 2.10 Four-quadrant use of the ATN function

range of the ATN function, and give correct polar angles automatically. In the second quadrant, with negative real part and positive imaginary part, ATN returns minus the angle C indicated in Figure 2.10, whereas the angle required in this case is $\pi + \text{ATN}(B/A)$, a positive angle. In the third quadrant, both parts are negative and ATN hence returns the positive angle C as indicated. The correct angle now is negative and given by $-(\pi - \text{ATN}(B/A))$.

(2) The modulus M of the polar form is calculated at line 130.

(3) At lines 140 and 150, tests are performed for the sign of the real part of the complex number. For positive real part, the angle or argument follows directly at line 160. The constant 57.296 converts radians to degrees.

(4) At line 190, the angle for a second quadrant complex number is evaluated. The corresponding calculation for a third quadrant number is performed at line 210.

(5) The special case of a pure imaginary number ($A = 0$) causes the program to branch to line 230. This condition is treated separately since any attempt to evaluate ATN(B/A) would require an inadmis-

sible divide by zero. From lines 230 to 270, the sign of **B** is tested to determine whether the angle is $+90°$, $-90°$, or $0°$ (if **B** is also zero). (6) Most of the complexity of this program results from the limitations of the standard ATN function supplied in BASIC. More advanced computer languages contain a four-quadrant version of the function, which would considerably simplify the resulting program.

Example 2.3: RTEMP: the effect of temperature on conductor performance

The effect of temperature on resistivity of metals, over a limited temperature range, may be represented by the equation:

$$\rho = \rho_o (1 + \alpha T) \qquad\qquad (2.30)$$

where ρ is resistivity at $T°C$, ρ_o is the value at $0°C$, and α is a constant termed the *temperature coefficient of resistance*. The resistivities of aluminium, copper, and silver at $0°C$ may be taken as 2.62×10^{-8}, 1.6×10^{-8}, and 1.47×10^{-8} Ωm respectively. Their corresponding α values are 0.0042, 0.0043, and 0.0040/°C. Write a program to calculate resistance, voltage drop and power loss variation with temperature between 20°C and 100°C for round wire of each material carrying a specified direct current.

Use the program to calculate values for 10m lengths of 1mm diameter round wire carrying a steady current of 5A, and comment on the significance of the results.

```
10 REM      RTEMP - Temperature effects on conductors
20 DATA 2.62E-8, 0.0042, 1.6E-8, 0.0043, 1.47E-8, 0.004
30 DIM S(2), A(2)
40 FOR I = 0 TO 2
50    READ S(I), A(I)
60    NEXT I
70 PRINT
80 PRINT
90 PRINT "Temperature effects on conductors"
100 PRINT "---------------------------------"
110 PRINT
120 PRINT
130 PRINT "Current flowing through wire (A)";
140 INPUT I1
150 PRINT
160 PRINT "Length of wire (in m)";
170 INPUT  L
180 PRINT
190 PRINT "Wire diameter (in mm)";
200 INPUT D
210 D = D/1000
220 K = L/(3.14159*D^2*0.25)
230 FOR I = 0 TO 2
240    PRINT
250    PRINT
```

```
260    IF I = 0 THEN PRINT TAB(28); "Aluminium"
270    IF I = 1 THEN PRINT TAB(29); "Copper"
280    IF I = 2 THEN PRINT TAB(29); "Silver"
290    PRINT
300    PRINT "    Temp (C)        R (Ohm)        ";
305    PRINT "Volt drop     Loss (W)"
310    FOR J = 20 TO 100 STEP 20
320      R = S(I)*(1 + A(I)*J)*K
330      V = R*I1
340      P = V*I1
350      PRINT TAB(7);J; TAB(16);R; TAB(30);V; TAB(44);P
360    NEXT J
370    NEXT I
380 PRINT
390 PRINT
400 PRINT "Another run (Y/N) ";
410 INPUT Q$
420 IF Q$ = "Y" THEN GOTO 110
430 STOP
440 END
```

```
Temperature effects on conductors
----------------------------------------

Current flowing through wire (A)?5

Length of wire (in m)?10

Wire diameter (in mm)?1
```

Aluminium

Temp (C)	R (Ohm)	Volt drop	Loss (W)
20	0.361610522	1.80805261	9.04026304
40	0.389632001	1.94816001	9.74080003
60	0.417653481	2.0882674	10.441337
80	0.445674961	2.2283748	11.141874
100	0.47369644	2.3684822	11.842411

Copper

Temp (C)	R (Ohm)	Volt drop	Loss (W)
20	0.22123829	1.10619145	5.53095725
40	0.238758081	1.1937904	5.96895202
60	0.256277872	1.28138936	6.4069468
80	0.273797663	1.36898831	6.84494157
100	0.291317454	1.45658727	7.28293634

Silver

Temp (C)	R (Ohm)	Volt drop	Loss (W)
20	0.202139681	1.0106984	5.05349202
40	0.21711299	1.08556495	5.42782476
60	0.2320863	1.1604315	5.8021575
80	0.24705961	1.23529805	6.17649024
100	0.262032919	1.3101646	6.55082298

```
Another run (Y/N) ?N

STOP at line 430
```

Program notes
(1) From lines 20 to 60, DATA and READ statements are employed to store resistivity and temperature coefficient values for the three materials in arrays S and A respectively. Note that array subscripts start from 0, and hence that S(2) consists of the three elements S(0), S(1), and S(2).

(2) At line 210, D in mm is converted to m before the factor K of length/cross-sectional-area for the particular wire dimensions is calculated at line 220.

(3) Lines 230 to 370 cover two nested FOR/NEXT loops. The first, using control variable I, covers labelling and table headings for each of the three materials. Control variable J is used in the second (inner) loop, and covers the five temperatures 20°, 40°, 60°, 80°, and 100°C. Note the use of indentation in printing the FOR/NEXT loops, which is intended to improve the readability of the program.

(4) Resistance, using Equation (2.30), is calculated for each material and temperature at line 320.

(5) In printing the values in the table at line 350, TAB statements are included to line up the values under their corresponding table headings.

(6) The specified run shows the linear variation in both voltage drop and power loss with temperature. Both variations have similar form, and the voltage drop characteristics are shown in Figure 2.11. For the 5A current in the specified conductors, a temperature of 100°C may easily be reached from an ambient temperature of 20°C. With increased voltage drops and losses over this range of about 30%, the

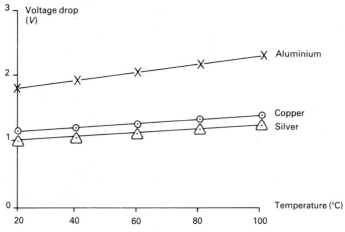

Figure 2.11 Voltage drop results from RTEMP

implication is that resistance variation with temperature is very significant, and must generally be allowed for.

Of the three materials, silver is the best conductor. However, its use is normally ruled out on cost grounds, and so the vast majority of electrical conductors are manufactured from copper. From time to time, however, the cost of copper has risen dramatically owing to supply difficulties. During these periods, aluminium conductors have become economically competitive, despite their larger bulk for a given resistance. Aluminium has the added advantage over copper of reduced weight for a given resistance because of its lower specific gravity.

Example 2.4: WAVES: characteristics of periodic waveforms

The significance of the r.m.s. value of a periodic waveform has been explained. Write a program to calculate the r.m.s. and half-cycle average values of a waveform specified at an even number of equal intervals over a half cycle. Calculate also the *form factor* of the waveform, defined as r.m.s./average. The integrals required should be evaluated using Simpson's Rule.

```
10 REM      WAVES - Analysis of periodic waveform
20 PRINT
30 PRINT
40 PRINT   "Analysis of periodic waveform"
50 PRINT   "-----------------------------"
60 PRINT
70 PRINT   "Periodic waveform data to be entered at EQUAL."
80 PRINT   "angular intervals over HALF cycle with EVEN"
90 PRINT   "number of intervals (e.g. at 1, 2, 3,"
95 PRINT   "5, 6, 9, 10, 15 deg intervals etc.)"·
100 PRINT
110 PRINT  "Ang interval (in deg) at which ";
115 PRINT  "data to be entered";
120 INPUT H
130 I = 180/H
140 IF INT(I) <> I THEN GOTO 630
150 IF (-1)^I < 0 THEN GOTO 630
160 PRINT
170 PRINT  "Enter magnitudes in following table"
180 PRINT  "at appropriate intervals when prompted:"
190 PRINT
200 PRINT  "(NOTE: For alternating waveforms, angle MUST"
210 PRINT  "       be measured from zero crossover point"
220 PRINT  "       to obtain correct values for"
225 PRINT  "       half-cycle average and form factor)"
230 PRINT
240 PRINT
250 PRINT  " Angle (deg)      Magnitude"
260 PRINT  " -----------      ---------"
270 PRINT
280 PRINT  "     0"; TAB(20);
```

```
290 INPUT V
300 A = 0
310 R = 0
320 A = A + V
330 R = R + V^2
340 FOR T = 1 TO I-1
350   PRINT TAB(5); T*H; TAB(20);
360   INPUT V
370   IF (-1)^(T+2) < 0 THEN A = A + 4*V
380   IF (-1)^(T+2) > 0 THEN A = A + 2*V
390   IF (-1)^(T+2) < 0 THEN R = R + 4*V^2
400   IF (-1)^(T+2) > 0 THEN R = R + 2*V^2
410   NEXT T
420 PRINT "       180"; TAB(20);
430 INPUT V
440 A = A + V
450 R = R + V^2
460 PRINT " -----------     -----------"
470 PRINT
480 PRINT
490 A = A/(3*I)
500 R = SQR(R/(3*I))
510 F = R/A
520 PRINT "Half-cycle average = "; A
530 PRINT
540 PRINT "     R.M.S. value = "; R
550 PRINT
560 PRINT "       Form factor = "; F
570 PRINT
580 PRINT
590 PRINT "Another run (Y/N) ";
600 INPUT Q$
610 IF Q$ = "Y" THEN GOTO 20
620 STOP
630 PRINT
640 PRINT "This angle does not give";
645 PRINT " an EVEN number of intervals -"
650 PRINT " Please try again"
660 GOTO 60
670 STOP
680 END
```

>RUN

Analysis of periodic waveform

Periodic waveform data to be entered at EQUAL
angular intervals over HALF cycle with EVEN
number of intervals (e.g. at 1, 2, 3,
5, 6, 9, 10, 15 deg intervals etc.)

Ang interval (in deg) at which data to be entered?15

Enter magnitudes in following table
at appropriate intervals when prompted:

(NOTE: For alternating waveforms, angle MUST

```
be measured from zero crossover point
to obtain correct values for
half-cycle average and form factor)


Angle (deg)          Magnitude
------------          ----------

    0                   ?0
   15                   ?.25882
   30                   ?.5
   45                   ?.70711
   60                   ?.86603
   75                   ?.96593
   90                   ?1
  105                   ?.96593
  120                   ?.86603
  135                   ?.70711
  150                   ?.5
  165                   ?.25882
  180                   ?0
------------          ----------


Half-cycle average = 0.636638889

     R.M.S. value = 0.707109466

     Form factor = 1.1106916


Another run (Y/N) ?Y


Analysis of periodic waveform
-----------------------------

Periodic waveform data to be entered at EQUAL
angular intervals over HALF cycle with EVEN
number of intervals (e.g. at 1, 2, 3,
5, 6, 9, 10, 15 deg intervals etc.)

Ang interval (in deg) at which data to be entered?20

This angle does not give an EVEN number of intervals -
 Please try again

Periodic waveform data to be entered at EQUAL
angular intervals over HALF cycle with EVEN
number of intervals (e.g. at 1, 2, 3,
5, 6, 9, 10, 15 deg intervals etc.)

Ang interval (in deg) at which data to be entered?9

Enter magnitudes in following table
at appropriate intervals when prompted:

(NOTE: For alternating waveforms, angle MUST
       be measured from zero crossover point
       to obtain correct values for
       half-cycle average and form factor)
```

```
Angle (deg)       Magnitude
------------      ---------

    0                ?0
    9                ?.1
   18                ?.2
   27                ?.3
   36                ?.4
   45                ?.5
   54                ?.6
   63                ?.7
   72                ?.8
   81                ?.9
   90                ?1
   99                ?.9
  108                ?.8
  117                ?.7
  126                ?.6
  135                ?.5
  144                ?.4
  153                ?.3
  162                ?,.2
  171                ?.1
  180                ?0
------------      ---------


Half-cycle average = 0.5

        R.M.S. value = 0.577350269

        Form factor = 1.15470054


Another run (Y/N) ?N

STOP at line 620
```

Program notes
(1) The reader unfamiliar with Simpson's rule for evaluating integrals is referred to the appropriate companion volume (Mason, 1983). To apply the method directly, the half cycle of the periodic waveform must have values specified at an even number of equal intervals. At line 130, the number of intervals is calculated from the angular interval input at line 120. The conditions that the number is an integer, and then an even integer, are checked at lines 140 and 150 respectively. Unless both conditions are satisfied, the user is required to enter the angular interval again via a suitable warning message.
(2) Lines 290 to 450 cover the summation of point values according to Simpson's Rule. The cumulative totals are held in A and R for calculating the average and the mean square respectively. The first and last points are treated individually at the beginning and end, and lines 370 to 400 process the intermediate even and odd values.

(3) The average and mean square values are evaluated at lines 490 and 500 by effectively dividing the integral by the total integration interval (180°).

(4) The first run processes a sine wave of peak value 1, input at 15° intervals. The values produced of $2/\pi$, $1/\sqrt{2}$, and 1.11 for average, r.m.s., and form factor may be verified by simple manual integrations in this case.

(5) The second run is begun by an inadmissible angular interval input of 20° (giving an odd number of intervals). The run proper then processes a symmetrical triangular waveform of peak value 1. The average value of 1/2 is obvious by inspection, and the r.m.s. value of $1/\sqrt{3}$ may again be checked by manual integration. Note that the form factor has increased over the sine wave case.

(6) Form factor is a useful figure of merit for a periodic waveform. It has a value of 1 for a waveform representing steady d.c. or a rectangular symmetrical a.c. waveform switching between equal positive and negative values. Increase in form factor above this minimum value is an indication of increased harmonic content in a waveform. Since harmonics give increased losses in electric circuits without (usually) any attendant benefits, it is generally beneficial from the power viewpoint for form-factor values to be kept as low as possible.

(7) The program may be used to analyse a half cycle of any symmetrical alternating waveform, or periodically varying, direct waveform. However, the need to integrate from a zero-crossover point for an alternating waveform should be self-evident.

(8) Simpson's Rule gives best results with smooth waveforms. Waveforms with discrete steps are preferably analysed using different techniques (see Problem (2.3)).

PROBLEMS

(2.1) The square root of a complex number is sometimes required in circuit calculations. Calculate expressions for the real and imaginary parts of the square root of a complex number input in its Cartesian form. Write a program to perform the calculations, and display the results. Check particularly that the program can square root negative real numbers correctly.

(2.2) Equations (2.14) and (2.15) may represent sine waves of voltage and current displaced by a general phase angle φ. Write a program to display values of these and also instantaneous power (vi) for ωt values between 0 and 2π at intervals of $\pi/24$. Input data should consist of r.m.s. values of voltage and current, and the phase angle φ.

Note particularly the form and frequency of the instantaneous power. What is its mean value when $\varphi = 0$ and $\varphi = \pi/2$?

(2.3) Some alternating electrical waveforms may be idealised as a sequence of rectangular steps. Write a program to calculate half-cycle average, r.m.s. value, and form factor for a general stepped waveform. Data required for each step is its height, and angular distance of its start from the zero crossover point. The number of steps in the half cycle should also be input, to enable the program to calculate the amount of data to be expected.

The integrations required to calculate the mean and mean square values of these rectilinear waveforms may be evaluated exactly by algebraic means. This much simplifies the programming required, compared to the general method used by WAVES.

(2.4) Produce a table of conductor resistance/unit length at 20°C and conductor weight/unit length, for copper and aluminium. Data should be calculated for conductor diameters of 1–5mm in 1mm steps. Take resistivity values to be 1.73×10^{-8} and 2.84×10^{-8} Ωm respectively, and corresponding specific gravities 8.9 and 2.7.

(2.5) The *dielectric strength* of an insulator is the maximum voltage gradient (V/m) it can withstand without sparking and subsequent breakdown occurring. It is required to design a 0.1 μF parallel-plate capacitor to have minimum possible plate area. The operating voltage is to be 2 kV. Dielectrics available are air, polystyrene, and mica. Relative permittivity values are 1, 2.7, and 6 respectively. Corresponding dielectric strengths may be taken as 5 kV/mm, 20 kV/mm and 200 kV/mm for the conditions considered. Write a program based on Equation (2.25) to calculate the plate area required in each case. Express the results in cm^2 and display the side length of each plate in cm, assuming the plates to be square.

What are the practical implications of the results?

Linear electric circuits

ESSENTIAL THEORY

3.1 Introduction

Linear circuit analysis is the determination of currents or voltages in networks, supplied by defined sources of electrical energy. These sources may be considered to be either of constant voltage or constant current, but any real source may be modelled by a single generator of either type. Linear circuit elements consist of resistance, inductance and capacitance as described in the previous chapter, together with mutual inductance which represents coupling between circuits via a magnetic field. For the purposes of this chapter, all these elements are assumed to have the required linearity property that their component values are constant, being unaffected by voltage, current, temperature, frequency etc.

Power is an important circuit consideration. Its calculation in d.c. circuits is straightforward, but in a.c. circuits its oscillatory nature and the general phase displacement between current and voltage introduce some complications.

3.2 Kirchhoff's voltage law and mesh current analysis

Two laws due to Kirchhoff form the basis of electric circuit theory. The voltage law may be expressed as

Around any closed path in a network, the algebraic sum of e.m.f.s and voltage drops is zero.

The use of the law in a d.c. circuit is illustrated in Figure 3.1. The e.m.f.s represent the active electrical sources in the circuit, and the voltage drops occur across each of the passive circuit elements. Note that the arrowheads on the potential difference arrows are placed at the higher potential end. Hence they are placed at the positive ends of the sources of e.m.f., and at the end at which current enters for the circuit elements. Application of Kirchhoff's law in a clockwise direction around the closed loop shown in Figure 3.1(a) gives:

$$E_1 - V_1 - E_2 - V_2 - V_3 + V_4 + V_5 = 0 \tag{3.1}$$

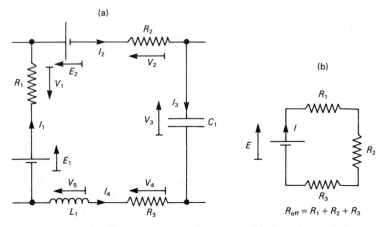

Figure 3.1(a) Kirchhoff's voltage law in a d.c. circuit   (b) Resistors in series

Note that the currents may be inserted in any direction in the branches of the network, but that the volt drop polarities in the circuit elements are then fixed by the chosen directions for the individual currents.

Using the voltage/current relationships developed in the previous chapter, Equation 3.1 may be rewritten as

$$E_1 - R_1I_1 - E_2 - R_2I_2 - 1/C_1\int_0^t I_3 dt + R_3I_4 + L_1\frac{dI_4}{dt} = 0 \qquad (3.2)$$

The presence of the integral sign associated with the capacitor indicates that the circuit's performance depends on the initial state of charge of the capacitor.

The closed path considered may represent part of a complete circuit. With similar equations representing the remaining closed paths in the circuit, sufficient independent simultaneous equations are obtainable to allow solution for all the currents present.

Solution of the simple single-loop circuit shown in Figure 3.1(b) gives

$$E = R_1I + R_2I + R_3I = (R_1 + R_2 + R_3)I \qquad (3.3)$$

which shows that the effective value of resistors in series in a branch is equal to the sum of the individual values.

The sinusoidal a.c. equivalents of these circuits are shown in Figure 3.2. Here, advantage may be taken of complex notation to represent inductance and capacitance by their corresponding reactances. Then, as before, application of Kirchhoff's voltage law gives

$$\mathbf{E}_1 - \mathbf{V}_1 - \mathbf{E}_2 - \mathbf{V}_2 - \mathbf{V}_3 + \mathbf{V}_4 + \mathbf{V}_5 = 0 \qquad (3.4)$$

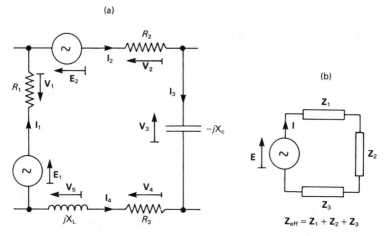

Figure 3.2(a) Kirchhoff's voltage law in an a.c. circuit   (b) impedances in series

where here the bold type indicates phasor notation, and hence that complex arithmetic is implied. Substitution for the voltage drops then gives

$$\mathbf{E}_1 - R_1\mathbf{I}_1 - \mathbf{E}_2 - R_2\mathbf{I}_2 + jX_c\mathbf{I}_3 + R_3\mathbf{I}_4 + jX_L\mathbf{I}_4 = 0 \qquad (3.5)$$

Note particularly the sign change introduced by X_c in Equation (3.5).

The values of X_c and X_L are fixed by the angular frequency of the supplies. With known supply values, and appropriate equations representing other loops in the rest of the circuit (not shown), solution for the currents could be obtained as before. It should be emphasised however, that this phasor technique only applies for sinusoidal supplies with multiple sources, when all sources operate at the same frequency.

The sinusoidal a.c. equivalent of resistances in series is impedances in series, and this situation is shown in Figure 3.2(b). From previous considerations it should be evident that the overall impedance of components in series is similarly the sum of the individual component impedances.

Before considering circuit solution in detail, a clear distinction must be made between *steady-state* and *transient* performance. When sources are initially switched into circuit, there is a (usually) short transient period before the circuit settles down to a steady condition. This is caused by inductance and capacitance, since inductance prevents current from instantaneously attaining its steady-state value, and capacitance similarly constrains the voltage. After the transient period, all voltages and currents are constant in a circuit with d.c. sources, which implies that inductances act as short-circuits ($V_L = 0$), and capacitors as open-circuits $(I_c = 0)$. Steady-state in the sinusoidally-supplied a.c. circuit has all voltages and currents varying periodically at the supply frequency. General circuit operation is represented by differential equations, and in mathematical terms, the steady-state solution corresponds to the *particular integral*, and the transient solution corresponds to the *complementary function*, obtained by setting the forcing functions (sources) to zero. The complete solution is the sum of the two, obtainable when the initial circuit conditions are known.

In most cases, the steady-state solution is of greater interest, the transient only being of importance for specialised considerations of fuse rating and circuit and device protection and formerly for timing purposes in electronic circuits. If attention is now restricted to steady-state alone, and hence any inductance or capacitance in the d.c. case, if represented in the circuit, is replaced by short-circuit and open-circuit respectively, then similar methods of analysis may be employed for both d.c. and sinusoidal a.c. circuits, with real arithmetic used for d.c. and complex arithmetic for a.c. Resistance in the d.c. circuit then corresponds to impedance in the a.c. circuit, and further thought leads to the fuller table of corresponding quantities below.

Feature	d.c. (R only)	a.c.
Voltage	V (a real number)	$\mathbf{V}$ (a complex number)
Current	I (a real number)	$\mathbf{I}$ (a complex number)
Ratio of above	Resistance R (a real number)	Impedance $\mathbf{Z}$ (a complex number)
Magnitude	Steady value	r.m.s. value (modulus of complex number)

A general method of circuit analysis, based on Kirchhoff's voltage law, and applicable to either of the above cases, may now be developed. It is explained by reference to the circuit shown in Figure 3.3, in which three closed paths may be readily identified. The circuit has been drawn in the form of a sinusoidal a.c. problem, with rectangles representing general impedances, but from the above table, the equivalent d.c. problem could be readily formulated.

The basis of the method is the assignment of a loop current (sometimes called a *Maxwell circulating current*) to each of the closed paths. The three loop currents, $\mathbf{I}_1$, $\mathbf{I}_2$, and $\mathbf{I}_3$ are indicated in the diagram. Each current is drawn in the same circulatory direction (clockwise in this case). Inspection of the circuit shows that the current through $\mathbf{Z}_A$ is $\mathbf{I}_1$, whereas the nett current through $\mathbf{Z}_B$ is $\mathbf{I}_1 - \mathbf{I}_2$. Currents through each of the other impedances may be similarly identified.

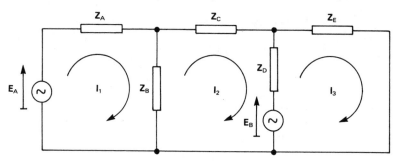

Figure 3.3 Circuit for mesh current analysis

Application of Kirchhoff's voltage law to each of the three loops in turn gives the following equations

$$E_A - Z_A I_1 - Z_B (I_1 - I_2) = 0 \qquad (3.6)$$
$$- E_B - Z_B (I_2 - I_1) - Z_c I_2 - Z_D (I_2 - I_3) = 0 \qquad (3.7)$$
$$E_B - Z_D (I_3 - I_2) - Z_E I_3 = 0 \qquad (3.8)$$

which can be readily solved for the loop currents using standard methods if E_A and E_B are known.

In order to develop a routine method of writing the loop equations, it is necessary to re-arrange the three simultaneous equations as follows

$$E_A = (Z_A + Z_B)I_1 \qquad\qquad - Z_B I_2 \qquad\qquad\qquad (3.9)$$
$$- E_B = \qquad - Z_B I_1 + (Z_B + Z_c + Z_D)I_2 \qquad - Z_D I_3 \qquad (3.10)$$
$$E_B = \qquad\qquad\qquad - Z_D I_2 + (Z_D + Z_E)I_3 \qquad (3.11)$$

Inspection of Equation 3.9 which concerns loop 1 shows the left-hand side to be the e.m.f. acting in the direction of the loop current. Consideration of the right hand side shows the loop current I_1 to be multiplied by $Z_A + Z_B$, the total impedance in loop 1. Z_B is also multiplied by I_2 because it is common to both loops 1 and 2. The negative sign for $Z_B I_2$ indicates that current 2 is in opposition to current 1, or that they pass through the common impedance in opposite directions. Equation 3.10 follows similarly. The left-hand side is $- E_B$ because that is the total e.m.f. in loop 2, and the negative sign indicates that it acts in the opposite direction to the loop current. I_2 is multiplied by the total impedance in loop 2, and I_1 and I_3 are multiplied by the impedances common, or mutual, with loops 1 and 3 respectively. Again, negative signs indicate that I_1 and I_3 pass through their respective mutual impedances in the opposite direction to the main loop current I_2. Similar arguments apply to Equation 3.11 and loop 3, but here the e.m.f. is positive because it acts in the same direction as its loop current.

At this point, it is expedient to introduce a double-subscript notation to indicate to which loops, of a possible two, a particular impedance belongs. So Z_{12} (or Z_{21}), indicates an impedance common to loops 1 and 2. Z_{11} is defined as the total impedance in loop 1, i.e. it also includes any impedance common to loop 1 and any other loop.

The equations are now most conveniently expressed in a general matrix form, and the reader unfamiliar with matrices is referred to the appropriate companion volume (Mason, 1984).

The general equations are

$$\begin{pmatrix} E_1 \\ E_2 \\ E_3 \end{pmatrix} = \begin{pmatrix} Z_{11} & -Z_{12} & -Z_{13} \\ -Z_{21} & Z_{22} & -Z_{23} \\ -Z_{31} & -Z_{32} & Z_{33} \end{pmatrix} \begin{pmatrix} I_1 \\ I_2 \\ I_3 \end{pmatrix} \tag{3.12}$$

where, in each case, the loop e.m.f. is the algebraic sum of the e.m.f.s acting in the same direction as the loop current.

Note that the circuit is completely general, allowing for any loop to be linked to any other, but full mutual linkage rarely happens in practice.

The advantage of this systematic approach is that it allows the loop equations to be written directly by inspection, in a routine way, prior to their solution for the loop currents using standard mathematical techniques.

The fact that $Z_{12} = Z_{21}$, etc. means that the impedance matrix is symmetrical about the main diagonal of total impedances. This property results in *reciprocity*, and the *reciprocity theorem* is

In any network of linear bilateral impedances containing a single voltage source, the ratio of source voltage to current measured in any branch is unchanged if the position of source and measuring point are interchanged.

A *bilateral* impedance has voltage/current properties which are independent of the direction in which it is connected into a network.

The ratio measured in the above theorem is termed the *transfer impedance* between the two points.

3.3 Inclusion of mutual inductance in networks

So far, the linear circuit elements considered have been limited to resistance, inductance and capacitance. There is one final element, or more precisely circuit property, which must be considered to completely define the behaviour of linear networks. This final property is *mutual inductance*.

In chapter 2, induced e.m.f. in a coil was related to its changing flux linkage by Faraday's law. There, the flux was assumed to be associated with the coil's own inductance, and its own current, but this restriction is unnecessary; e.m.f. may be similarly induced in one coil by changing flux linkage, where the flux originates from another coil. This effect is mutual inductance. The single-coil effect is more strictly termed *self inductance*.

Mutual inductance links circuit loops magnetically, or by a common flux, as opposed to the electric linkage, by a common current, of the common or mutual loop elements previously considered. This magnetic linkage is illustrated in Figure 3.4. Two coils linked by a

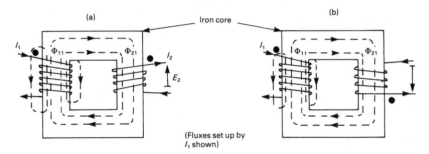

Figure 3.4 Polarity of induced e.m.f. in magnetically-coupled coils

common flux are shown in Figure 3.4(a). They are shown wound on opposite sides of a common rectangular core so that the magnetic coupling can be clearly seen. Current I_1 in coil 1 sets up a flux Φ_{11}, such that the self inductance of the coil is given by

$$L_{11} = N_1 \Phi_{11}/I_1 \tag{3.13}$$

as before. The double-subscript notation is now introduced to distinguish between self and mutual inductance.

A proportion of the total flux Φ_{11} also links coil 2. This flux is denoted by Φ_{21}, and since it links both coils, it is called the mutual flux. The mutual inductance between the two coils, M_{21}, then follows as

$$M_{21} = N_2 \Phi_{21}/I_1 \tag{3.14}$$

being the flux linkage of coil 2 set up by the current in coil 1. From an identical argument as was previously applied to self inductance, the voltage induced in coil 2 follows directly as

$$E_2 = M_{21}\frac{dI_1}{dt} \tag{3.15}$$

With the winding directions visible, as in Figure 3.4(a), the relative directions of the voltages in the two windings may be determined using Lenz' law. With current I_1 in the direction shown, flux is driven in a clockwise direction around the core. From Lenz' law, voltage must be induced in coil 2 in such a direction that any resulting current will set up an opposing flux. Hence, with I_1 increasing, the top of coil 2 must be at a positive potential with respect to the bottom, with current flowing outwards. Conversely, in Figure 3.4(b) with coil 2 wound in the opposite sense, the polarity and current directions are reversed.

The magnetic coupling may similarly be established by a current flowing in coil 2 producing a flux linking coil 1 (Φ_{12}) via the corresponding mutual inductance M_{12}. Since the magnetic circuit for the mutual flux is independent of which coil sets it up, it can be shown that

$$M_{12} = M_{21} = M \tag{3.16}$$

and the subscripts may be dropped in this case. However, the principle applies to any number of circuits, and if more than two are considered, then the double-subscript notation must be retained to avoid ambiguity.

M may be related to the self inductances of the two coils since

$$\begin{aligned} L_{11}L_{22} &= (N_1\Phi_{11}/I_1)\,(N_2\Phi_{22}/I_2) \\ &= (N_1\Phi_{22}/I_2)\,(N_2\Phi_{11}/I_1) \end{aligned} \tag{3.17}$$

and

$$M = \sqrt{((N_1\Phi_{12}/I_2)\,(N_2\Phi_{21}/I_1))} \tag{3.18}$$

If a constant k is introduced (<1), such that $k^2 = \{\Phi_{12}/\Phi_{22}\}\cdot\{\Phi_{21}/\Phi_{11}\}$ then

$$M = k\sqrt{(L_{11}L_{22})} \tag{3.19}$$

Since k defines how closely the coils are linked magnetically, it is called the *coefficient of coupling*.

In the general case, currents flow in both coils, and the overall circuit equations for the two coils, neglecting winding resistance, are

$$V_1 = L_{11}dI_1/dt \pm MdI_2/dt \tag{3.20}$$
$$V_2 = L_{22}dI_2/dt \pm MdI_1/dt \tag{3.21}$$

where the signs of the mutual voltages depend on the respective winding directions. For sinusoidal supplies, complex notation may again be used to simplify the equations to

$$\mathbf{V}_1 = j\omega(L_{11}\,\mathbf{I}_1 \pm M\mathbf{I}_2) \tag{3.22}$$
$$\mathbf{V}_2 = j\omega(L_{22}\,\mathbf{I}_2 \pm M\mathbf{I}_1) \tag{3.23}$$

It is necessary to introduce a convention to determine the polarity of the mutual voltages in the above equations. Hence, for circuit analysis purposes, a dot is assigned to a pair of coil ends which have the same natural polarity. In other words, if current is instantaneously increasing through one of the coils at the dotted end, the dotted end of the other coil has induced positive polarity at that instant, with the tendency to increase current out of the coil at that end. Using this rule, both tops or both bottoms of the coils in Figure 3.4(a) would be dotted, whereas one top and the other bottom would be dotted of the coils shown in Figure 3.4(b). For mesh current analysis, the coils of Figure 3.4(a) would be represented as indicated in Figure 3.5(a).

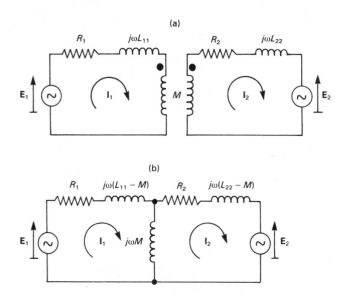

Figure 3.5(a) dot notation for mutual inductance   (b) electrical equivalent circuit

The natural current directions, with one current entering at a dotted end and the other leaving give subtractive fluxes and hence negative signs for the mutual voltages, since the overall induced voltage in either circuit must be reduced. Conversely, both currents entering or both leaving at the dotted ends give additive fluxes and hence positive signs for the mutual voltages and increased overall induced voltages. The mesh currents of Figure 3.5(a) correspond to the former condition, and so the coupled circuit equations reduce to

$$\mathbf{E}_1 = (R_1 + j\omega L_{11})\mathbf{I}_1 - j\omega M\mathbf{I}_2 \tag{3.24}$$
$$-\mathbf{E}_2 = -j\omega M\mathbf{I}_1 + (R_2 + j\omega L_{22})\mathbf{I}_2 \tag{3.25}$$

(Note the presence of the negative sign for the source voltage in Equation 3.25.)

For network-analysis purposes, it is more convenient to replace the magnetic coupling with an equivalent electrical coupling. For this example, this is achieved by the circuit represented in Figure 3.5(b). The mesh equations for the loops of this circuit are the above two equations, and the circuit may be solved directly without further consideration of the magnetic coupling involved. It should be noted, however, that coupled circuits with widely differing self-inductances may not have physically realisable equivalent electrical networks, since one of the inductances may be negative. Nevertheless, the circuit accurately models the behaviour of the magnetic coupling.

The principles involved are well illustrated by the examples of mutually-coupled coils in series. Figure 3.6(a) represents the coils

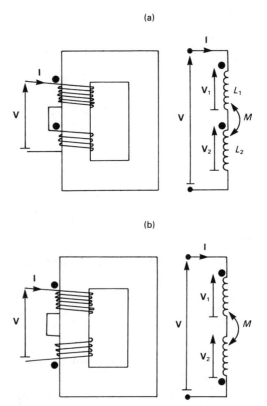

Figure 3.6 Series connected mutually coupled coils (a) cumulative coupling (b) differential coupling

connected cumulatively or in series addition. With sinusoidal excitation the circuit equations are

$$V_1 = j\omega(L_1 + M)I \tag{3.26}$$
$$V_2 = j\omega(L_2 + M)I \tag{3.27}$$

The total voltage is the sum of the two components and hence the overall effective inductance is $L_1 + L_2 + 2M$. The converse case of the coils connected differentially or in series opposition is represented in Figure 3.6(b). The effective inductance is now evidently $L_1 + L_2 - 2M$.

These two connections can form the basis of a technique for mutual inductance measurement, and the differential connection of two closely-coupled identical air-cored coils forms the basis for the non-inductive rheostat used in electrical laboratories.

3.4 Kirchhoff's current law and node voltage analysis

Circuit analysis in terms of node potentials may be developed along similar lines to those previously employed for mesh currents, but it is necessary first to discuss sources, and in particular to describe the current source.

The sources so far considered have been voltage sources, representing typically a battery in the d.c. case, or a sinusoidal, constant-frequency supply derived from the National Grid in the a.c. case. The voltage developed is assumed constant, but any source must have some resistance (or impedance in the a.c. case), resulting in an essentially linear fall in available voltage as the current supplied is increased. The voltage source and its voltage/current characteristic are hence as indicated in Figure 3.7. In fact, all the circuits considered here could be reduced to the simple form of Figure 3.7 by making use of *Thévenin's Theorem* which is

The current in any impedance connected to two terminals of a network is unchanged if the network is replaced by a single voltage

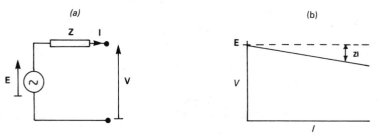

Figure 3.7(a) Thévenin voltage source   (b) operating characteristic

source in series with a single impedance; the voltage being the open circuit voltage appearing at the terminals, and the impedance being the impedance looking back into the network, with voltage sources replaced by their internal impedances.

Use of this theorem enables any network, ultimately, to be reduced to a single voltage source, and a single impedance.

Current sources may be introduced via *Norton's Theorem* which is complementary to Thévenin's theorem and states

The current in any impedance connected to two terminals of a network is unchanged if the network is replaced by a single current source in parallel with a single impedance; the current being the current flowing through the two terminals when they are short circuited, and the impedance being the impedance looking back into the network, with current sources being replaced by their internal impedances.

The consequence of the two theorems is that there must be a direct equivalence between real voltage and current sources, and the Norton current source equivalent to the Thévenin voltage source of Figure 3.7 is shown in Figure 3.8. Although a real source may be

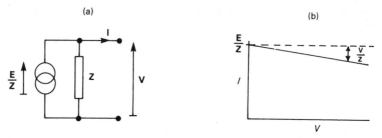

Figure 3.8(a) Norton current source   (b) operating characteristic

represented in either way, one which develops approximately constant terminal voltage irrespective of current drawn is normally considered as a voltage source. Conversely, one which delivers virtually constant current regardless of the load connected to its terminals is regarded as a current source. Voltage sources are more familiar because they are more commonly encountered; current sources not usually being met until semiconductor circuits have been studied.

The quality of the source depends on the value of the source impedance. A good voltage source requires a low source impedance, and a good current source requires a high one. It should be noted that perfect voltage and current sources, with zero and infinite internal impedances respectively, cannot be transformed. The Norton equivalent of a perfect voltage source would act as a short circuit, and the Thévenin equivalent of a perfect current source would act as an open circuit.

To perform a nodal analysis on a network, it is first essential to express all sources as current generators. Analysis is further simplified if impedance is expressed in its reciprocal form, termed *admittance* (*Y*). The resistive part of this has already been encountered in the form of conductance (*G*), but note that the conductance, being strictly the real part of the admittance of a complex circuit, is not in general equal to the reciprocal of the resistance of the circuit. The reactive, or imaginary, part in terms of complex notation, is termed *susceptance* (*B*). The units of admittance and susceptance must be the same as those for conductance, the Siemen (S). Hence the corresponding complex forms are

$$\mathbf{V} = \mathbf{Z}\mathbf{I} \text{ where } \mathbf{Z} = R + jX \tag{3.28}$$
$$\mathbf{I} = \mathbf{Y}\mathbf{V} \text{ where } \mathbf{Y} = G + jB \tag{3.29}$$

A circuit in the correct form for nodal analysis is represented in Figure 3.9. The *principal nodes* or *junctions* in the circuit are connections between three or more circuit elements. One of these principal nodes must be chosen as the *reference node*, and the aim of the method is to calculate the potentials of all other nodes with respect to this node. In Figure 3.9, the bottom node has been selected as ref-

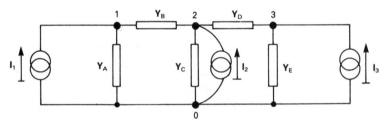

Figure 3.9 Circuit for nodal analysis

erence and labelled 0, and the three other junctions have been labelled 1, 2, and 3. The three potentials to be calculated are strictly V_{10}, V_{20}, and V_{30} respectively, but since node 0 is common, the double-subscript notation may be dropped and the voltages called V_1, V_2, and V_3 without ambiguity. Equations are developed by applying *Kirchhoff's current law* at junctions 1, 2, and 3 in turn. This law states simply that

At any junction in an electrical network, the algebraic sum of the currents is zero.

If current is considered as a flow of electrons, Kirchhoff's current law may be regarded as a principle of conservation of charge. Its application at nodes 1 to 3 in the network gives

$$I_1 - V_1 Y_A - (V_1 - V_2) Y_B = 0 \tag{3.30}$$
$$I_2 - (V_2 - V_1) Y_B - V_2 Y_C - (V_2 - V_3) Y_D = 0 \tag{3.31}$$
$$I_3 - (V_3 - V_2) Y_D - V_3 Y_E = 0 \tag{3.32}$$

where current flow into each node has been considered positive. The equations may be readily re-arranged in matrix form as

$$\begin{pmatrix} I_1 \\ I_2 \\ I_3 \end{pmatrix} = \begin{pmatrix} (Y_A + Y_B) & -Y_B & \\ -Y_B & (Y_B + Y_C + Y_D) & -Y_D \\ & -Y_D & (Y_D + Y_E) \end{pmatrix} \begin{pmatrix} V_1 \\ V_2 \\ V_3 \end{pmatrix} \tag{3.33}$$

As for mesh currents, these equations may be solved for node voltages, using real arithmetic in the d.c. case where admittance is restricted to conductance only, or complex arithmetic for the sinusoidal a.c. case. Again, the equations may be written in a completely general matrix form by employing double-subscript notation to give

$$\begin{pmatrix} I_1 \\ I_2 \\ I_3 \end{pmatrix} = \begin{pmatrix} Y_{11} & -Y_{12} & -Y_{13} \\ -Y_{21} & Y_{22} & -Y_{23} \\ -Y_{31} & -Y_{32} & Y_{33} \end{pmatrix} \begin{pmatrix} V_1 \\ V_2 \\ V_3 \end{pmatrix} \tag{3.34}$$

It is, in fact, even simpler to write down the node equations by inspection than it is to produce the loop equations. The self-admittances are the algebraic sums of all the admittances connected to the appropriate node. The mutual admittance between two nodes is the total admittance between the two, with the sign reversed. The currents are the algebraic sums of the current sources feeding into the appropriate node. Hence, a nett outward current at a node is negative.

The similarity between the mesh current and node voltage equations should be noted. This mathematical correspondence between differing representations of quantities in the same engineering discipline is called *duality*. In loop and nodal analysis, current is the dual of voltage, and impedance the dual of admittance.

The reciprocity theorem stated in Section 3.2 has a corresponding form in terms of current sources and node potentials. In this case, the ratio obtained would be the *transfer admittance*.

The choice between loop or nodal analysis for networks is clearly arbitrary, since either must give the same final solution. However, loop analysis is probably more generally known and used because of the unfamiliarity of current sources. This is unfortunate because nodes in circuits are frequently easier to identify than loops, and, with practice, nodal analysis can often produce a quicker solution. A simple example of this is the case of a current source feeding several parallel admittances, as represented in Figure 3.10. Here only two junctions are present, giving the one nodal equation

$$I = (Y_A + Y_B + Y_C + Y_D)V_1 \tag{3.35}$$

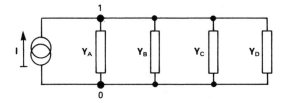

Figure 3.10 Admittances in parallel

showing that the effective admittance of a parallel combination is the sum of the individual admittance values. This implies that parallel resistances (in d.c.) or impedances must be added reciprocally. In principle, solution of this circuit by mesh current analysis would require conversion of the current source to an equivalent voltage source, conversion from admittance to impedance for the four elements, and the simultaneous solution of three equations for the mesh currents.

This is obviously an extreme example, but it shows that it is advisable to become familiar with both methods, so that the most appropriate one may be applied in each case.

3.5 Power

The power associated with an element or source is given directly by the product of potential difference between the ends of the device (or voltage across it) and the current flowing through it. The unit of power is the Watt (W).

In the steady-state d.c. case, its calculation is quite straightforward with both voltage and current having constant values. The power supplied by a source is given by the product of terminal voltage and supplied current, with the current emerging from the positive end. The power supplied to a resistance is given by the corresponding product, but with the current entering at the positive end. The power in the resistance is lost, or dissipated, as heat. This situation is illus-

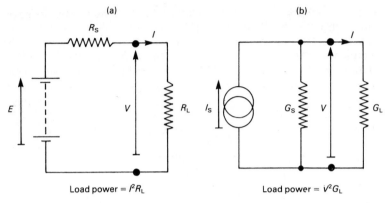

Figure 3.11 d.c. power (a) real voltage source   (b) real current source

trated in Figures 3.11(a) and (b) for a real voltage source and a real current source respectively.

Power in a.c. circuits requires more careful consideration, since the product of alternating voltage and current results in power which is varying in time. If the discussion is restricted to sinusoidal conditions, the general condition may be represented as indicated in Figure 3.12. If voltage is given by $\sqrt{2}V \sin \omega t$ and current by $\sqrt{2}I \sin (\omega t - \varphi)$, then the instantaneous power P follows as

$$P = 2VI \sin \omega t \sin (\omega t - \varphi) \tag{3.36}$$

The mean power is normally of more interest, since it indicates the overall direction of energy flow in a circuit. To extract the mean power from the above equation it is re-expressed in the alternative form

$$P = VI(\cos \varphi - \cos (2\omega t - \varphi)) \tag{3.37}$$

Since the second term varies cosinusoidally with time, its mean value is zero. Hence, the mean power $\bar{P}$ is given by

$$\bar{P} = VI \cos \varphi \tag{3.38}$$

The mean and instantaneous powers are clearly indicated in Figure 3.12(a). Note that the instantaneous power oscillates at twice the supply frequency.

The product of r.m.s. voltage and current must be multiplied by the additional factor $\cos\varphi$ to give the mean power. For this reason, $\cos\varphi$ is termed the *power factor*. Evidently, the mean power is VI when $\varphi = 0$, i.e. when V and I are coincident or in phase. Conversely, the mean power is zero when $\varphi = \pi/2$, or when V and I are in quadrature.

It should be recalled that pure inductors and pure capacitors have quadrature voltage-current relationships under sinusoidal conditions. Hence they have no mean power associated with them. This is because the power oscillates back and forth in these devices. Energy is stored in the magnetic or electric field respectively on the positive half cycles, and returned to the supply on the negative half cycles. Hence the nett energy transfer is zero.

Evidently from the phasor diagram of Figure 3.12(b), the active, or power, component of current is $I \cos \varphi$. The corresponding reactive, or quadrature, component is $I \sin \varphi$, and the quantity $VI \sin \varphi$ is termed the *reactive power Q*. Since it is a product of voltage and current, but does not represent power, it is given the special unit, var, which indicates 'volt-ampere-reactive'. The quantity VI is called the *apparent power*, but again it is not true power and so it is given another unit, VA, and the symbol S. Hence

$$S^2 = \bar{P}^2 + Q^2 \tag{3.39}$$

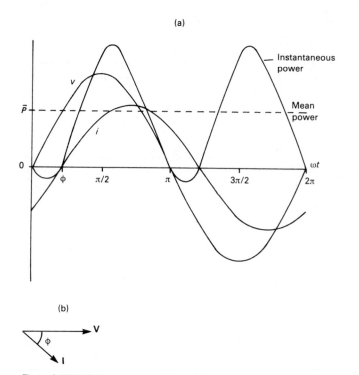

(a)

(b)

Figure 3.12(a) Voltage, current, and power waveforms in sinusoidal a.c. circuit (b) phasor diagram

and a power diagram, corresponding to the phasor diagram of Figure 3.12(b), may be drawn relating mean power, apparent power, and reactive power.

It is useful to give a sign to reactive power to indicate whether current is lagging or leading voltage. The convention adopted is that the sign of Q is the same as the sign of the reactance producing it. Hence

Inductive load, $\mathbf{Z} = R + jX$, has positive Q, and is said to *absorb Q*. Current *lags* the voltage by a positive phase angle φ, and the power factor is said to be lagging.

Capacitive load, $\mathbf{Z} = R - jX$, has negative Q, and is said to *supply Q*. Current *leads* the voltage by a negative phase angle φ, and the power factor is said to be leading.

These power relationships may be concisely stated using complex notation. Since the complex product involves angular addition, and the complex conjugate $\mathbf{Z}^*$ of complex number $\mathbf{Z}$ has unchanged modulus, but reversed sign of argument, then from Figure 3.12(b)

$$\mathbf{VI}^* = VI \angle \varphi = VI \cos \varphi + jVI \sin \varphi \tag{3.40}$$

Evidently, the modulus is S, the real part $\bar{P}$, and the imaginary part Q. Note that the alternative complex product, $\mathbf{V}^*\mathbf{I}$ gives the correct magnitude but the incorrect sign for Q.

Since $\bar{P} = VI \cos \varphi$, power factor and current are inversely related for fixed values of mean power and supply voltage. Since conductor cross-sectional area and the size of electrical generation and distribution equipment is related to current demand, it is essential to keep power factor high in electrical power plant to minimise capital cost. Most electrical equipment operates at a lagging power factor, since electric motors, which make up the bulk of electrical loads, are essentially magnetically-coupled coils and inductive in nature. The power factor seen by the supply may be increased by the connection of a capacitor of appropriate value in parallel with the inductive power plant, as indicated in Figure 3.13. Note that the overall current drawn from the supply is reduced, but that the plant current and plant power are unaffected.

This is the principle of *power factor correction*.

3.6 Power transfer and efficiency

Perfect electrical sources transfer all their generated or stored energy to a connected electrical load and are 100% efficient. However, this is not true for real sources which have internal impedance and, in gen-

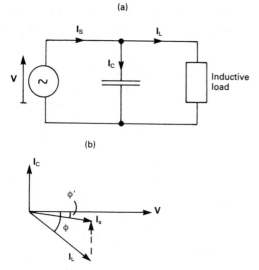

Load power factor cos φ
Corrected supply power factor cos φ'

Figure 3.13 Power factor correction (a) circuit (b) phasor diagram

eral, internal power loss. Complex notation may be used to analyse the general sinusoidal a.c. case with a voltage generator of e.m.f. $\mathbf{E}$ and internal impedance $\mathbf{Z}_s$ connected to an electrical load $\mathbf{Z}_L$. Then since $\mathbf{E} = (\mathbf{Z}_s + \mathbf{Z}_L)\mathbf{I}$ and the load voltage $\mathbf{V} = \mathbf{Z}_L\mathbf{I}$

$$\mathbf{VI^*} = \frac{\mathbf{E}\mathbf{Z}_L\mathbf{E^*}}{(\mathbf{Z}_s + \mathbf{Z}_L)(\mathbf{Z}_s + \mathbf{Z}_L)^*}$$

$$= \frac{\mathbf{Z}_L|\mathbf{E}|^2}{|\mathbf{Z}_s + \mathbf{Z}_L|^2} \tag{3.41}$$

where $|\ \ |$ indicates taking the modulus of the complex number. $\bar{P}$ is given by the real part of $\mathbf{VI^*}$. If $\mathbf{Z}_s = R_s + jX_s$ and $\mathbf{Z}_L = R_L + jX_L$, where either X could in principle be capacitive, and hence negative, then

$$\bar{P} = \frac{E^2 R_L}{(R_s + R_L)^2 + (X_s + X_L)^2} \tag{3.42}$$

Equation 3.42 may be successively differentiated with respect to R_L and X_L and the results set to zero to determine conditions for maximum power transfer to the load. The conditions are that $R_L = R_s$, and $X_L = -X_s$ which proves the *Maximum Power-Transfer Theorem* for sinusoidal a.c. conditions, which may be stated as

Maximum power will be transmitted from a source to a variable load when the load impedance is made equal to the complex conjugate of the source impedance

The corresponding form for resistances and d.c. sources has the condition of equal source and load resistance.

It should be noted that, at the maximum power transfer point, half the power is dissipated in the source, giving only 50% power efficiency, and only half the source e.m.f. appears across the load. A dichotomy appears here between the requirements of electronic circuits, and those of power supply and transmission circuits. In the former, maximum power transfer from source to load is a common requirement, and load matching using the above principles is frequently employed. In the latter, high transfer efficiency and relatively constant system voltage are the main criteria, and so load impedances are employed which are very many times higher than the source impedance. In fact, maximum power transfer could not be achieved in conventional power supply circuits without catastrophic overheating of the generators and supply lines.

3.7 Bibliography

MASON, J. C., *BASIC Matrix Methods*, Butterworths, London (1984)

WORKED EXAMPLES

Example 3.1: SOURCE: a.c. source impedance by measurement

The characteristics of a sinusoidal a.c. voltage source may be determined from measurements of voltage and current at its terminals. Develop the equations for source resistance and source reactance in terms of open-circuit voltage, and the currents flowing when two dissimilar (assumed pure) known resistances are connected as loads. Hence write a program to process the results of practical measurements.

```
10 REM      SOURCE - Source impedance calculation
20 PRINT
30 PRINT
40 PRINT "Source impedance calculation"
50 PRINT "----------------------------"
60 PRINT
70 PRINT "(Sinusoidal a.c. voltage source assumed)"
80 PRINT
90 PRINT
100 PRINT "Measured r.m.s. open-circuit voltage";
110 INPUT E
120 PRINT
130 PRINT "Value of 1st load resistance (Ohms)";
140 INPUT R1
150 PRINT
```

```
160 PRINT "1st measured r.m.s. current (Amps)";
170 INPUT I1
180 PRINT
190 PRINT "Value of 2nd load resistance (Ohms)";
200 INPUT R2
210 IF R2 = R1 GOTO 420
220 PRINT
230 PRINT "2nd measured r.m.s. current (Amps)";
240 INPUT I2
250 IF (R1-R2)/(I2-I1) < 0 GOTO 450
260 A = (E/I1)^2
270 B = (E/I2)^2
280 C = (R2-R1)*(R2+R1)
290 D = 2*(R1-R2)
300 R = (A-B+C)/D
310 F = (R+R1)^2
320 X = SQR(A-F)
330 PRINT
340 PRINT
350 PRINT "Source impedance is ";R;" +/- j";X;" Ohm"
360 PRINT
370 PRINT
380 PRINT "Another calculation (Y/N) ";
390 INPUT Q$
400 IF Q$ = "Y" GOTO 80
410 STOP
420 PRINT
430 PRINT "Load resistances must be dissimilar"
435 PRINT " - please try again"
440 GOTO 80
450 PRINT
460 PRINT "Measured values inconsistent"
465 PRINT " - please try again"
470 GOTO 80
480 STOP
490 END
```

```
SOURCE IMPEDANCE CALCULATION
----------------------------

(Sinusoidal a.c. voltage source assumed)


Measured r.m.s. open-circuit voltage?240

Value of 1st load resistance (Ohms)?10

1st measured r.m.s. current (Amps.)?22.5

Value of 2nd load resistance (Ohms)?20

2nd measured r.m.s. current (Amps.)?11.7


Source impedance is 0.349901378 +/- j2.58017814 Ohm


Another calculation (Y/N)?N

STOP at line 410
```

Program notes
(1) With open-circuit voltage E, source resistance R_s and source reactance X, the current I flowing when load resistance R is connected is given in magnitude by

$$I = E/\sqrt{((R_s + R)^2 + X^2)} \qquad (3.43)$$

Hence, if currents I_1 and I_2 flow when load resistances R_1 and R_2 respectively are connected, then solution of the resulting equations gives

$$R_s = \frac{(E/I_1)^2 - (E/I_2)^2 + R_2^2 - R_1^2}{2(R_1 - R_2)} \qquad (3.44)$$

$$X = \sqrt{((E/I_1)^2 - (R_s + R_1)^2)} \qquad (3.45)$$

These equations are evaluated from lines 260 to 320 of the program.
(2) Two simple checks are incorporated in this program to ensure that data entered is sensible. At line 210, a check is made that the two resistance values input are not identical. If they are identical, the program branches to a suitable error message at line 430, and then back to the beginning of the input data section (line 80), so that the user can start again. At line 250, the second check is made that the higher current corresponds to the lower load resistance. Failure of this test again causes a program branch to an error message (line 460), and subsequent return to line 80.

This procedure is called *data validation*. Its use adds considerably to the length and complexity of programs, and it is not possible to cover all possible misuses of programs, however much effort is devoted to it. Data validation is of greatest importance in large programs, or ones which take a long time to run, where incorrect input data may not be obvious from the results obtained, or excessive amounts of expensive computer time may be wasted. When any program of significant size or complexity is developed, an economic decision must be made as to how much data validation to incorporate.
(3) The test method cannot distinguish between inductive and capacitive source reactance. Hence the result, printed at line 350, incorporates '$+/-$' preceding the reactive part.

Example 3.2: PARRES: effect of frequency on characteristics of parallel a.c. circuit

A circuit commonly encountered consists of a capacitor in parallel

PARRES: effect of frequency on characteristics of parallel a.c. circuit

with a coil (inductance and resistance in series). Write a program to
read values of the three components, and to calculate admittance
values for the capacitor, coil, and overall circuit, as a function of
excitation frequency.

```
10 REM       PARRES - Effect of f on parallel a.c. circuit
20 PRINT
30 PRINT
40 PRINT "Parallel a.c. circuit"
50 PRINT "----------------------"
60 PRINT
70 PRINT "(C in parallel with R and L)"
80 PRINT
90 PRINT
100 PRINT "Value of C (in microFarads)";
110 INPUT C
120 PRINT
130 PRINT "Coil inductance (in milliHenries)";
140 INPUT L
150 PRINT
160 PRINT "Coil resistance (Ohms)";
170 INPUT R
180 C = C*1E-6
190 L = L*1E-3
200 PRINT
210 PRINT
220 PRINT "Freq (Hz) Yc (S)       Ycoil (S) ";
225 PRINT "         Ymod (S)   Angle"
230 PRINT
240 FOR M = 0 TO 6
250    FOR N = 1 TO 7 STEP 3
260      F = 10^M*N
270      W = 2.0*3.14159*F
280      G1 = 0
290      B1 = W*C
300      D2 = R^2 + (W*L)^2
310      G2 = R/D2
320      B2 = -W*L/D2
330      G = G1 + G2
340      B = B1 + B2
350      Y = SQR(G^2 + B^2)
360      ANG = ATN(B/G)*180/3.14159
370      PRINT TAB(2);F; TAB(8);"j"; B1; TAB(19);G2;
380      PRINT TAB(29);"j"; B2; TAB(41);Y; TAB(51);ANG
390    NEXT N
400    NEXT M
410 PRINT
420 PRINT
430 PRINT "Another run (Y/N) ";
440 INPUT Q$
450 IF Q$ = "Y" THEN GOTO 20
460 STOP
470 END
```

```
>RUN


Parallel a.c. circuit
----------------------

(C in parallel with R and L)


Value of C (in microFarads)?.1

Coil inductance (in milliHenries)?2.5

Coil resistance (Ohms)?1
```

Freq (Hz)	Yc (S)	Ycoil (S)		Ymod (S)	Angle
1	j6.28E-7	1	j-1.57E-2	1	-0.9
4	j2.51E-6	0.996	j-6.26E-2	0.998	-3.6
7	j4.4E-6	0.988	j-0.109	0.994	-6.27
10	j6.28E-6	0.976	j-0.153	0.988	-8.93
40	j2.51E-5	0.717	j-0.45	0.847	-32.1
70	j4.4E-5	0.453	j-0.498	0.673	-47.7
100	j6.28E-5	0.288	j-0.453	0.537	-57.5
400	j2.51E-4	2.47E-2	j-0.155	0.157	-80.9
700	j4.4E-4	8.2E-3	j-9.02E-2	9.01E-2	-84.8
1E3	j6.28E-4	4.04E-3	j-6.34E-2	6.29E-2	-86.3
4E3	j2.51E-3	2.53E-4	j-1.59E-2	1.34E-2	-88.9
7E3	j4.4E-3	8.27E-5	j-9.09E-3	4.7E-3	-89
1E4	j6.28E-3	4.05E-5	j-6.37E-3	9.22E-5	-63.9
4E4	j2.51E-2	2.53E-6	j-1.59E-3	2.35E-2	90
7E4	j4.4E-2	8.27E-7	j-9.09E-4	4.31E-2	90
1E5	j6.28E-2	4.05E-7	j-6.37E-4	6.22E-2	90
4E5	j0.251	2.53E-8	j-1.59E-4	0.251	90
7E5	j0.44	8.27E-9	j-9.09E-5	0.44	90
1E6	j0.628	4.05E-9	j-6.37E-5	0.628	90
4E6	j2.51	2.53E-10	j-1.59E-5	2.51	90
7E6	j4.4	8.27E-11	j-9.09E-6	4.4	90

```
Another run (Y/N) ?Y


Parallel a.c. circuit
----------------------

(C in parallel with R and L)


Value of C (in microFarads)?.1

Coil inductance (in milliHenries)?2.5

Coil resistance (Ohms)?200
```

PARRES: effect of frequency on characteristics of parallel a.c. circuit

Freq (Hz)	Yc (S)	Ycoil	(S)	Ymod (S)	Angle
1	j6.28E-7	5E-3	j-3.93E-7	5E-3	2.7E-3
4	j2.51E-6	5E-3	j-1.57E-6	5E-3	1.08E-2
7	j4.4E-6	5E-3	j-2.75E-6	5E-3	1.89E-2
10	j6.28E-6	5E-3	j-3.93E-6	5E-3	2.7E-2
40	j2.51E-5	5E-3	j-1.57E-5	5E-3	0.108
70	j4.4E-5	5E-3	j-2.75E-5	5E-3	0.189
100	j6.28E-5	5E-3	j-3.93E-5	5E-3	0.27
400	j2.51E-4	5E-3	j-1.57E-4	5E-3	1.08
700	j4.4E-4	4.98E-3	j-2.74E-4	4.99E-3	1.9
1E3	j6.28E-4	4.97E-3	j-3.9E-4	4.98E-3	2.74
4E3	j2.51E-3	4.55E-3	j-1.43E-3	4.68E-3	13.4
7E3	j4.4E-3	3.84E-3	j-2.11E-3	4.47E-3	30.8
1E4	j6.28E-3	3.09E-3	j-2.43E-3	4.94E-3	51.3
4E4	j2.51E-2	4.6E-4	j-1.45E-3	2.37E-2	88.9
7E4	j4.4E-2	1.6E-4	j-8.8E-4	4.31E-2	89.8
1E5	j6.28E-2	7.98E-5	j-6.26E-4	6.22E-2	89.9
4E5	j0.251	5.06E-6	j-1.59E-4	0.251	90
7E5	j0.44	1.65E-6	j-9.09E-5	0.44	90
1E6	j0.628	8.1E-7	j-6.37E-5	0.628	90
4E6	j2.51	5.07E-8	j-1.59E-5	2.51	90
7E6	j4.4	1.65E-8	j-9.09E-6	4.4	90

Another run (Y/N) ?N

STOP at line 460

Program notes
(1) The circuit concerned is shown in Figure 3.14(a). After the component values have been read in, the capacitance and inductance values are converted to Farads and Henries respectively at lines 180 and 190.
(2) The frequency values for admittance calculation are varied in an essentially logarithmic manner by a double FOR-NEXT loop, spanning lines 240–400. The outer loop variable M covers the decade or power of 10 of the frequency, and the inner loop variable N steps through values of 1, 4 and 7 in each decade.
(3) For each frequency value, admittance values for the capacitor and coil are calculated in terms of their real and imaginary parts conductance G, and susceptance B. Note that the capacitor conductance G1, is zero, and that, being the reciprocal of reactance, capacitive susceptance is positive, and inductive susceptance negative. The admittancies of the series components representing the coil add reciprocally, and this operation is carried out at lines 300 to 320.
(4) The total circuit admittance is expressed in polar form. The modulus is calculated at line 350, and the angle, converted to degrees, appears at line 360.

(a)

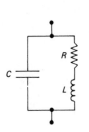

(b)

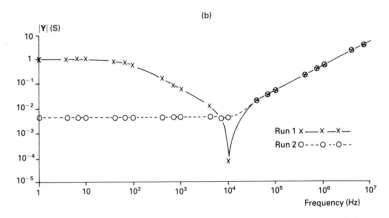

Figure 3.14 Parallel resonance (a) circuit   (b) admittance results from PARRES

(5) The first run shows a sharp admittance minimum at a frequency of 10 kHz, which may be clearly seen on the log–log graph of the results shown in Figure 3.14(b). The capacitor susceptance increases with frequency, whereas that of the coil falls. At 10 kHz, the two exactly cancel, leaving only the coil conductance as the total circuit admittance. A phenomenon of this type is termed *resonance*, and the frequency of 10 kHz at which the voltage and current are in phase is termed the *resonant frequency* of this *parallel resonant circuit*. Resonance is exploited in frequency-selective or tuning circuits.
(6) Simple algebra shows the resonant frequency f_0 to be given by

$$f_0 = \frac{1}{2\pi\sqrt{(LC)}}\Bigg/\sqrt{\left(1 - \frac{R^2 C}{L}\right)} \tag{3.46}$$

which may be checked to give a value of 10 kHz for the component values 0.1 μF, 2.5 mH and 1 Ω used in run 1. It should be evident from the equation that increasing C, or R, or reducing L eventually results in no real value of f_0. This is demonstrated in run 2, where R has been increased to $200\,\Omega$, and no admittance minimum is evident.

(7) Note that the admittance value at high frequency is identical for each run. This is because with increasing frequency the capacitive susceptance eventually becomes dominant.

Example 3.3: NODE: node voltage analysis

Develop a program to perform a node voltage analysis following the method developed in Section 3.4. The program should cater for a maximum of 10 principal nodes, including the reference node.

```
10 REM      NODE - Node voltage analysis
20 DIM I1(11), I2(11), G(11,11), B(11,11), V1(11)
30 DIM V2(11), G1(11), B1(11)
40 FOR I = 0 TO 11
50    I1(I) = 0
60    I2(I) = 0
70    G1(I) = 0
80    B1(I) = 0
90    V1(I) = 0
100   V2(I) = 0
110   FOR J = 0 TO 11
120      G(I,J) = 0
130      B(I,J) = 0
140   NEXT J
150  NEXT I
160 PRINT
170 PRINT
180 PRINT "Node voltage analysis using complex notation"
190 PRINT "------------------------------------------------"
200 PRINT
210 PRINT "Identify the principal nodes in your network"
220 PRINT "and enter total number (max allowed 10) :-";
230 INPUT N
240 IF N > 10 GOTO 1080
250 PRINT
260 PRINT "Select one node as your REFERENCE, and label"
270 PRINT "the others from 1 to "; N-1
280 PRINT
290 FOR I = 1 TO N-1
300    PRINT
310    PRINT "Total conductance connected to node "; I;
320    INPUT G(I-1,I-1)
330    PRINT
340    PRINT "Total susceptance connected to node "; I;
350    INPUT B(I-1,I-1)
360    IF I = N-1 GOTO 510
370    FOR J = I+1 TO N-1
380       PRINT
390       PRINT "Total conductance connected ";
395       PRINT "between node "; I
400       PRINT "and node "; J;
410       INPUT G(I-1,J-1)
420       G(I-1,J-1) = -G(I-1,J-1)
430       G(J-1,I-1) = G(I-1,J-1)
```

```
440      PRINT
450      PRINT "Total susceptance connected ";
455      PRINT "between node "; I
460      PRINT "and node "; J;
470      INPUT B(I-1,J-1)
480      B(I-1,J-1) = -B(I-1,J-1)
490      B(J-1,I-1) = B(I-1,J-1)
500      NEXT J
510    NEXT I
520 PRINT
530 PRINT
540 FOR I = 1 TO N-1
550    PRINT "REAL part of total source current ";
555    PRINT "IN to node "; I;
560    INPUT I1(I-1)
570    PRINT "IMAGINARY part of total source current ";
575    PRINT "IN to node "; I;
580    INPUT I2(I-1)
590    PRINT
600    NEXT I
610 FOR I = 1 TO N-2
620    G(I,N-1) = 0
630    B(I,N-1) = 0
640    NEXT I
650 G(0,N-1) = 1
660 B(0,N-1) = 0
670 FOR N1 = 0 TO N-2
680    FOR J = 0 TO N-2
690      D1 = G(0,0)^2 + B(0,0)^2
700      G(N-1,J) = (G(0,J+1)*G(0,0) + B(0,J+1)*B(0,0))/D1
710      B(N-1,J) = (B(0,J+1)*G(0,0) - G(0,J+1)*B(0,0))/D1
720      NEXT J
730    FOR I = 1 TO N-2
740      G1(I) = G(I,0)
750      B1(I) = B(I,0)
760      NEXT I
770    FOR J = 0 TO N-2
780      FOR I = 1 TO N-2
790        G(I,J) = G1(I)*G(N-1,J) - B1(I)*B(N-1,J)
800        B(I,J) = B1(I)*G(N-1,J) + G1(I)*B(N-1,J)
810        G(I,J) = G(I,J+1) - G(I,J)
820        B(I,J) = B(I,J+1) - B(I,J)
830        NEXT I
840      NEXT J
850    FOR J = 0 TO N-2
860      FOR I = 0 TO N-2
870        G(I,J) = G(I+1,J)
880        B(I,J) = B(I+1,J)
890        NEXT I
900      NEXT J
910    NEXT N1
920 FOR I = 0 TO N-2
930    FOR J = 0 TO N-2
940      V1(I) = V1(I) + G(I,J)*I1(J) - B(I,J)*I2(J)
950      V2(I) = V2(I) + B(I,J)*I1(J) + G(I,J)*I2(J)
960      NEXT J
970    NEXT I
980 PRINT
990 PRINT
1000 PRINT "Node      Potential(V)           ";
```

```
1005 PRINT "Real Part        Imag Part"
1010 PRINT "----    -------------          ";
1015 PRINT "----------        ---------"
1020 PRINT
1030 FOR I = 0 TO N-2
1040    PRINT TAB(1); I+1; TAB(10); SQR(V1(I)^2 + V2(I)^2);
1050    PRINT TAB(30); V1(I); TAB(47); V2(I)
1060    NEXT I
1070 STOP
1080 PRINT
1090 PRINT "*** Max allowed number of nodes is ten"
1100 PRINT " - Please try again ***"
1110 GOTO 160
1120 END
```

```
Node-voltage analysis using complex notation
--------------------------------------------

Identify the principal nodes in your network
and enter the total number (max allowed 10) :-?4

Select one node as your REFERENCE, and label
the others from 1 to 3


Total conductance connected to node 1?3

Total susceptance connected to node 1?1

Total conductance connected between node 1
and node 2?2

Total susceptance connected between node 1
and node 2?1

Total conductance connected between node 1
and node 3?0

Total susceptance connected between node 1
and node 3?0

Total conductance connected to node 2?6

Total susceptance connected to node 2?3

Total conductance connected between node 2
and node 3?3

Total susceptance connected between node 2
and node 3?0

Total conductance connected to node 3?4

Total susceptance connected to node 3?1


REAL part of total source current IN to node 1?10
IMAGINARY part of total source current IN to node 1?0
```

```
REAL part of total source current IN to node 2?5
IMAGINARY part of total source current IN to node 2?0

REAL part of total source current IN to node 3?5
IMAGINARY part of total source current IN to node 3?5
```

Node	Potential (V)	Real Part	Imag Part
1	6.38	5.93	-2.36
2	4.56	3.83	-2.48
3	4.04	3.73	-1.55

```
STOP at line 1070
```

Node-voltage analysis using complex notation

```
Identify the principal nodes in your network
and enter the total number (max allowed 10) :-?4

Select one node as your REFERENCE, and label
the others from 1 to 3


Total conductance connected to node 1?3

Total susceptance connected to node 1?0

Total conductance connected between node 1
and node 2?2

Total susceptance connected between node 1
and node 2?0

Total conductance connected between node 1
and node 3?0

Total susceptance connected between node 1
and node 3?0

Total conductance connected to node 2?6

Total susceptance connected to node 2?0

Total conductance connected between node 2
and node 3?3

Total susceptance connected between node 2
and node 3?0

Total conductance connected to node 3?4

Total susceptance connected to node 3?0


REAL part of total source current IN to node 1?10
IMAGINARY part of total source current IN to node 1?0
```

```
REAL part of total source current IN to node 2?5
IMAGINARY part of total source current IN to node 2?0

REAL part of total source current IN to node 3?5
IMAGINARY part of total source current IN to node 3?0
```

Node	Potential (V)	Real Part	Imag Part
1	7.59	7.59	0
2	6.38	6.38	0
3	6.03	6.03	0

```
STOP at line 1070
```

Program notes
(1) The program processes complex numbers exclusively in terms of real and imaginary parts. One-dimensional arrays are set up to accommodate real and imaginary parts of current sources (I1 and I2) and node potentials (V1 and V2). Two-dimensional arrays G and B are to hold initially the real parts (conductance) and imaginary parts (susceptance) of the admittance matrix, in the form of Equation (3.34), but in general with an increased number of components.

(2) All arrays are initially zero filled (lines 40–150). In many computers, this is the default condition, but it is not a universal one. Since zero initial conditions for array-element values is often implicitly assumed, it is good programming practice to make this self-evident.

(3) The program prompts the user to correctly identify and label principal nodes, and enter the total number, checking that this does not exceed 10.

(4) For each numbered node in sequence, starting at 1, the self and mutual admittances are input via a double FOR-NEXT loop between lines 300 and 510. The self values are first input, followed by the mutual values between the current node and each higher-numbered node. At each stage, the sign of any mutual conductance or susceptance is immediately reversed, as dictated by the theory. The symmetry of the matrix about its main diagonal enables the values of the remaining mutual elements to be assigned (lines 430 and 490).

(5) Current sources are input at each node in turn.

(6) The mathematical manipulation is performed between lines 610 and 910. The program performs a matrix inversion using complex arithmetic on the admittance matrix. It is suggested that the student takes this part of the program as read, but the reader interested in matrix inversion techniques is again referred to the appropriate companion volume (Mason, 1984). In essence, the matrix inversion converts the admittance matrix into the dual impedance matrix. The

extra arrays G1 and B1 are employed in this process, and after the inversion, G and B hold the real and imaginary parts of the impedance matrix, resistance and reactance respectively. The Equation (3.34) form has been converted to the form of Equation (3.12), where the Es are now the node potentials.

(7) Lines 920 to 970 cover the evaluation of the individual node potentials from the Equation (3.12) form. The real and imaginary parts of the complex products are obtained in the manner developed in Example 2.1. The total node potentials then follow from the sums of the appropriate IZ products at each node.

(8) The two sample runs solve a circuit of the form of Figure 3.9. The first is an a.c. problem, and the second illustrates how a d.c. circuit may be modelled by setting all imaginary inputs to zero.

PROBLEMS

(3.1) For a.c. circuit analysis purposes, a real inductance is often represented by an inductive reactance in series with a resistance. It may be equally represented by different values of the two elements in parallel. Develop the equations relating the two forms, and write a program to perform the conversion in both directions.

(3.2) Write a program to cover the general conversion of an impedance to an admittance and *vice versa*. Input should be in the form of real and imaginary parts. Alternative possible approaches are either to develop Cartesian equations and equate real and imaginary parts, or to change to polar forms and to calculate a complex reciprocal. The program should be able to handle successfully the special cases of pure resistance (or conductance) and pure reactance (or susceptance).

(3.3) Conversion between the two forms of circuit shown in Figure 3.15 can often effect considerable simplification in network analysis. Develop the equations relating R_A, R_B, and R_C in Figure 3.15(a) with R_1, R_2, and R_3 in Figure 3.15(b). For equivalence, the voltage/current relationships must be the same in each case. The two configurations are known as Δ or π and Y or T respectively. Hence write a program to perform $Y - \Delta$ *(or $T - \pi$) transformation*, and *vice versa*.

(3.4) Extend **(3.3)** to cover the more general case of impedance in place of resistance.

(3.5) Develop a program to perform a mesh current analysis for three loops, containing a combination of d.c. voltage sources and resistances only. The technique is outlined in Section 3.2. The program should prompt the user to first assign and draw loop currents. For each loop, the total source voltage is required, together with its polarity (i.e. sign) with respect to the loop current. Total (self) loop re-

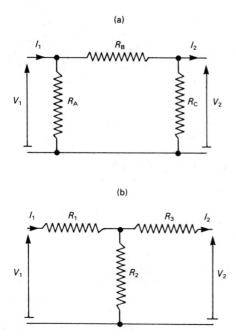

Figure 3.15(a) △ (π) network   (b) Y (T) network

sistance is then required. Finally, mutual resistances between the current loop and higher numbered loop(s) must be input. The resulting equations may be solved using Cramer's rule or Gauss elimination (refer to Mason, 1984, if necessary).

(3.6) Extend **(3.5)** to solve a 3-loop steady-state a.c. circuit. This requires replacement of real d.c. sources by complex a.c. ones, resistance by impedance, and real arithmetic by complex arithmetic.

(3.7) The mutual inductance between two coupled coils is to be determined. Write a program to perform the derivation from current measurements, with the coils connected in series cumulatively and differentially, and connected to a sinusoidal a.c. supply. Coil resistance may be assumed negligible in comparison with the inductive reactances involved. Input data required is supply voltage and frequency, and the values of the two currents.

(3.8) When the resistance of the two coils is significant, a wattmeter becomes necessary to calculate mutual inductance in the previous problem. Modify the program written for **(3.7)** to cater for this. Additional input required is the power reading corresponding to one of the currents.

(3.9) An electric circuit consists of resistance, inductance, and capacitance connected in series. Write a program, on the lines of PARRES, to display values of inductive and capacitive reactance, and overall circuit impedance, as a function of excitation frequency. Hence investigate the properties of a *series resonant circuit*. Component values should be read as input data.

(3.10) Write a program to convert between a Thévenin voltage source, and a Norton current source, and *vice versa*. The circuits are given in Figures 3.7 and 3.8 respectively. The complex form should be used, to cover both the sinusoidal a.c. and the d.c. case. The Thévenin form should be open-circuit voltage and series source impedance, and the complementary Norton form short-circuit current and parallel source admittance.

(3.11) An electrical load is connected to a sinusoidal a.c. voltage source of fixed frequency. A wattmeter and ammeter monitor supply power and current respectively. A capacitor is to be connected in parallel with the load to improve the power factor, which may be assumed to be lagging. Write a program to display uncorrected values of mean power, reactive power, apparent power, and power factór. These may be derived from input values of supply voltage and frequency, in addition to the initial instrument readings. Then extend the program to calculate and display the above quantities, and also supply current, for any value of correction capacitor input by the user.

(3.12) Develop a program to display current, output power, and efficiency for a real d.c. voltage source connected to a variable resistive load. The stepping technique used in PARRES should be used to vary load resistance from (say) $10R_s$ down to $0.1R_s$, where R_s is the source resistance. In this case, the output should be displayed at each $1/10$ decade interval. The results should demonstrate the maximum power transfer theorem for a resistive load.

(3.13) Extend the previous problem to cover the fixed-frequency a.c. case and complex impedance. Input data now becomes r.m.s. open-circuit voltage (a real number), and source impedance in terms of real and imaginary parts. Output should be extended to include power factor. Each run should now cover both variation in load resistance and load reactance. Load reactance is assigned the opposite sign to that of the source reactance. The load resistance should first be varied as in the previous example, with load reactance held at the same magnitude as the source reactance. Reactance should then be varied in a similar manner, with load resistance held equal to the source resistance. Note the power factor at the maximum power transfer point.

Magnetic circuits

ESSENTIAL THEORY

4.1 Introduction

Some basic aspects of magnetic circuit theory have been presented earlier in connection with inductance in electric circuits. A fuller consideration is necessary to facilitate understanding of subsequent work on transformers and electromechanical energy conversion.

In this chapter, the laws, quantities, and units of magnetic circuits are developed with the help of an idealised model of a simple circuit. There is an obvious analogy between electric and magnetic circuits, and this is explored. This leads to the concept of the magnetic equivalent circuit, which may be used to analyse more complex devices containing ferromagnetic material. The analogy does have significant limitations however, and these are pointed out.

Both conducting and ferromagnetic materials which experience a variable flux linkage are subject to power loss. These losses are often detrimental, and methods of estimating their magnitude and reducing their effect are considered.

Permanent magnets are widely used as magnetic-flux sources, particularly in small devices, and the analysis of simple permanent-magnet circuits is discussed.

4.2 Magnetic circuit parameters and laws

Figure 4.1 shows a magnetic circuit consisting of a C-shaped piece of iron with a multi-turn current-carrying coil wound around it. The current sets up a magnetic flux which completes continuous loops through the iron via the air region between its ends. Such a well-defined area conveying flux a short distance between iron regions is called an *airgap*.

Before attempting to calculate the performance of the magnetic circuit, it is important to appreciate that two effects are present which somewhat complicate the issue. The first is flux leakage, which results in less flux crossing the airgap than passing through the coil. This is caused by flux taking alternative leakage paths through the air in preference to the airgap, some of which are indicated in Figure 4.1. The effect may be approximately allowed for by the definition of a leakage coefficient k_l given by

$$k_l = \frac{\text{flux through coil}}{\text{flux through gap}} \tag{4.1}$$

Leakage flux is often very considerable, and leakage coefficients in excess of 2 are not uncommonly encountered. Careful circuit arrangement, endeavouring to keep flux paths through iron short, and to place the coil as close to the airgap as possible, is required to minimise the effect.

The second effect, normally much less significant unless the airgap is very long, is flux fringing at the gap. This is the tendency of the magnetic flux lines to bulge outwards at the edges of the gap, effectively increasing both the average flux path length across the gap, and also the effective gap cross section. This phenomenon is also indicated in Figure 4.1. It may be allowed for by an appropriate factor, but in practice its effect may often be neglected within the limits of calculation accuracy.

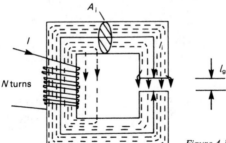

Figure 4.1 Magnetic circuit with airgap

In considerations of inductance in Chapter 2, it was explained that magnetic flux levels depend on current magnitudes and numbers of turns in coils. The product of these two quantities is termed *magnetomotive force* (m.m.f.), for reasons which are explained later. The unit of m.m.f. is strictly A, but it is commonly stated as AT to indicate that turns are also involved in addition to current. The symbol for m.m.f. is F.

The effect of the m.m.f. depends on the length of magnetic circuit over which it acts. The m.m.f./unit length is defined as *magnetic field strength*, which is given the symbol **H**, and has the unit A/m or AT/m. **H** is a vector quantity, having the same direction as the magnetic flux set up by the associated m.m.f.

Magnetic flux density, with symbol **B**, is a measure of how closely magnetic flux lines are packed together. It is defined as flux/unit cross-sectional area and is also a vector quantity. The unit of **B** is evidently Wb/m^2 which is given the name Tesla (T).

For the purposes of magnetic-circuit calculations, materials may be classified into one or two categories, depending on their relationship between magnetic field strength and magnetic flux density. Members of the first category include vacuum, air, and most other materials, excluding iron. For these, there is a linear relationship between **H** and **B** which is expressed by the equation

$$\mathbf{B} = \mu_o \mathbf{H} \tag{4.2}$$

The constant of proportionality, μ_o, is called the *permeability of free space*. From the definition of inductance, its unit must be H/m and the value of μ_o follows from the definition of the Ampère as $4\pi \times 10^{-7}$ H/m.

Members of the second category include iron, most steels, and a few other materials, and for this reason they are termed *ferromagnetics*. They are characterised by highly non-linear relationships between **B** and **H**. Hence, for these materials

$$\mathbf{B} = \mu_o \mu_r \mathbf{H} \tag{4.3}$$

μ_r is termed the *relative permeability* of the ferromagnetic material, and is evidently dimensionless. Its value varies typically from some hundreds to several thousands.

Similarity may be noted here between permeability in magnetic circuits, and the permittivity encountered in connection with capacitance in Chapter 2. However, it is vital to remember that, unlike relative permittivity, relative permeability is not, in general, constant.

The high values of permeability of ferromagnetics are the reason for their use in supplying or guiding flux to where it is required in magnetic circuits. Their non-linear **B**–**H** relationships are commonly represented graphically, and typical **B**–**H** curves for a selection of ferromagnetic materials are shown in Figure 4.2. In transformers (described in the next chapter) magnetic flux alternates, and **B** and **H** vectors vary in magnitude and sign, but are directed along one axis only. Advantage may be taken of this, to enhance the magnetic properties along this axis at the expense of properties on the orthogonal axes. *Grain-oriented transformer iron* is such a material, and the **B**–**H** characteristic shown is for the preferred axis. Conversely in electrical machines, the **B** and **H** vectors in general rotate, and little benefit is gained by using steels with magnetically-directional properties. Non-

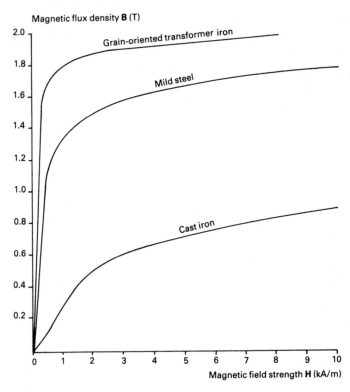

Figure 4.2 Typical **B–H** curves for ferromagnetic materials

directional steels have **B–H** characteristics of the same general order as that of mild steel shown in Figure 4.2.

Two laws are used to relate the various quantities in the magnetic circuit. The first is a simple flux continuity condition, whereby if flux density varies along a flux path, the condition that the flux does not change dictates that the product of flux density and cross-sectional area must be constant. The second is the relationship between magnetic field strength and m.m.f. which may be stated as

Around any closed path, the m.m.f. is equal to the sum of the products of individual path lengths and magnetic field strengths.

This is a specific form of the *Ampère circuital law*, the general form of which is beyond the scope of this book.

4.3 Solution of simple magnetic circuit

Two equations may now be derived which specify the operation of the magnetic circuit shown in Figure 4.1

$$k_l \mathbf{B_g} A_g = \mathbf{B_i} A_i = \Phi \tag{4.4}$$

This is the flux continuity condition for the flux set up by the core, with an allowance for leakage flux. In this case, $A_g = A_i$.

$$\mathbf{H_g} l_g + \mathbf{H_i} l_i = NI = F \tag{4.5}$$

Two forms of calculation are then necessary, depending on whether flux or current values are known in the circuit.

If flux is known, either in the coil or the gap, then the flux densities may be obtained from Equation (4.4), assuming that the physical dimensions of the coil are known. The values of $\mathbf{H}$ are then required for substitution in Equation (4.5). For the airgap $\mathbf{H_g} = \mathbf{B_g}/\mu_o$ and may be calculated directly. Due to the magnetic non-linearity of the iron, $\mathbf{H_i}$ is normally obtained from $\mathbf{B_i}$ via a **B–H** curve or a table of **B–H** values supplied for the particular material. Less often, a value for relative permeability μ_r may be given, which must be assumed valid for the conditions considered. Use of Equation (4.5) then gives the required value of m.m.f. F, and finally the coil current required to set up the specified flux, if the number of turns in the coil is known.

The alternative problem of flux calculation when current and hence m.m.f. is known is slightly more complex. This arises because a graphical approach must be used to evaluate the subdivision of m.m.f. between the iron and the airgap. One relationship between $\mathbf{B_i}$ and $\mathbf{H_i}$ is given by the **B–H** curve for the iron (or a corresponding table). Another must be derived from Equations (4.4) and (4.5). Since $\mathbf{H_g} = \mathbf{B_g}/\mu_o$, Equation (4.4) gives

$$\mathbf{H_g} = \mathbf{B_i} A_i/(k_l \mu_o A_g) \tag{4.6}$$

Substitution for $\mathbf{H_g}$ in Equation (4.5) then gives

$$\mathbf{B_i} A_i l_g/(k_l \mu_o A_g) + \mathbf{H_i} l_i = NI \tag{4.7}$$

which is the second relationship required between $\mathbf{B_i}$ and $\mathbf{H_i}$. Equation (4.7) is a straight line which may be plotted on the **B–H** curve of the iron as indicated in Figure 4.3. The intercepts on the **B** and **H** axes may be readily calculated, and the crossing point of the two curves gives the operating point $(\mathbf{B_i}, \mathbf{H_i})$ of the iron. Flux values then follow directly from Equation (4.4).

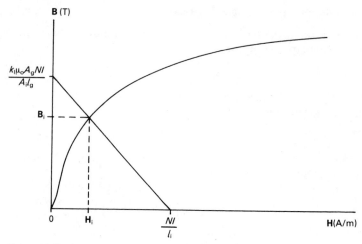

Figure 4.3 Derivation of flux from known m.m.f. in simple magnetic circuit

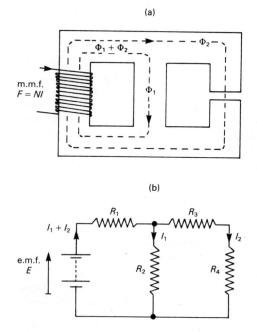

Figure 4.4 Analogous magnetic and electric circuits (a) magnetic circuit
(b) equivalent electric circuit

4.4 Analogy between magnetic and electric circuits

In the analysis of more complex magnetic circuits, it may be helpful to think in terms of an equivalent electric circuit, since relatively complex electrical networks are familiarly encountered, and a close analogy may be drawn between electrical and magnetic quantities.

Figure 4.4(a) illustrates a three-limbed magnetic circuit with an excitation coil wound on one outer limb, and an airgap inserted in the other outer limb. The effects of magnetic leakage and fringing are ignored in this case. Clearly, flux driven by the coil m.m.f. will divide between the parallel magnetic paths of the circuit in a manner depending on the various 'magnetic resistances'. Hence, it is reasonable to draw an equivalent electrical network as shown in Figure 4.4(b). Here, R_1 represents the resistance of the left-hand limb and horizontal sections carrying the total coil flux. R_2 represents the resistance of the central limb. Flux in the right-hand limb passes through an iron section broken by an airgap, and resistances R_3 and R_4 represent the iron and the airgap respectively.

To make effective use of the analogy for calculation purposes, it is necessary to determine an expression for 'magnetic resistance'. Magnetic flux Φ is evidently analogous to electric current, the magnitude of each being dependent on its appropriate motive force. The expression required is consequently F/Φ for a magnetic circuit element, corresponding to E/I for the electric circuit. But from the magnetic circuit laws of the previous section

$$F/\Phi = \mathbf{H}l/(\mathbf{B}A)$$

Since $\mathbf{B} = \mu\mathbf{H}$ in the magnetic circuit, the magnetic equivalent of resistance, called *reluctance* ($\mathscr{R}$) simplifies to

$$\mathscr{R} = l/(\mu A) \tag{4.8}$$

With μ having the unit H/m, the unit of $\mathscr{R}$ is evidently H^{-1}.

The analogy is closely maintained, since both electrical resistance and magnetic reluctance are proportional to component (or path) length, and inversely proportional to cross-sectional area. Reference to Section 2.4 shows that permeability μ is analogous to conductivity σ. As for electrical circuits, it is convenient to consider also the inverse of resistance. The magnetic equivalent of conductance G is *permeance* Λ and so

$$\Lambda = \mu A/l \, \mathrm{H} \tag{4.9}$$

Owing to the mathematical correspondence, operations of combining series reluctances by addition, and parallel permeances by addition are, in principle, permissible. However, having completed the analogy, it is important to bear in mind its limitations, based on important differences between electric and magnetic circuits.

Firstly, a current in an electric circuit represents a flow of electrons, whereas there is no flow associated with magnetic flux. Secondly, leakage currents in an electric circuit are normally negligible, whereas leakage fluxes in magnetic circuits are rarely so. This is essentially because there is typically a conductivity variation between conductors and insulators of 20 orders of magnitude, whereas permeability variation between ferromagnetic material and air is rarely greater than 10 000. In terms of the analogy, non-magnetic materials are not particularly good 'magnetic insulators'. The final, and most important difference, is that resistance in electrical circuits is commonly fairly constant. Reluctance, however, is rarely anything like constant in magnetic circuits containing iron, making them non-linear. The direct methods of analysis described in Chapter 3 are only applicable to linear circuits, where resistance (or impedance) is constant. Hence, only rarely will the simple analytical techniques developed for electric circuits be directly applicable to magnetic circuits. More commonly, graphical techniques using **B–H** curves, or iterative techniques which require repeated use to steadily improve accuracy of solution, must be employed. Despite these differences, the analogy provides a useful aid to understanding how magnetic circuits perform, and it is particularly useful when effects of circuit modification on performance must be estimated.

4.5 Iron losses

Losses occur whenever conducting materials experience a time-varying flux linkage. These are generally termed *iron losses* since they are commonly associated with the ferromagnetic parts of electrical machines. However, they are not confined to circuits containing iron.

When investigating iron losses, two distinct mechanisms emerge. The first is *hysteresis loss,* arising from the time-varying magnetisation of ferromagnetic material. This loss is clearly unique to magnetic iron. The second mechanism is the e.m.f. induced by the variable flux linkage, and its consequent current flow in conductors which results in an I^2R loss. The currents are termed *eddy currents* and the loss is known as *eddy-current loss.*

As previously indicated, in a.c. devices the iron experiences a magnetic field strength which either alternates or rotates. The flux density responds to this variation in a manner related to the appropriate **B–H** curve. However, the flux density variation 'lags' the magnetic field strength variation, and the **B–H** curves of Figure 4.2 become **B–H** loops under steady-state a.c. conditions. These loops

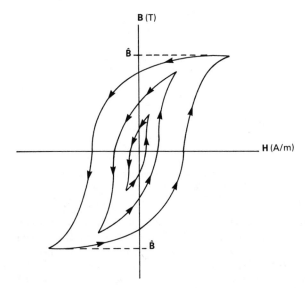

Figure 4.5 Typical hysteresis loops for a ferromagnetic material

are termed hysteresis loops, and forms for a typical material are sketched in Figure 4.5. It can be readily demonstrated that the loop represents a loss of energy and further proved that the energy loss in unit volume of the material for each cycle of magnetisation is given by the loop area. For individual materials, the loop area increases with peak loop flux density value $\hat{\mathbf{B}}$. It has been found by experiment that loop area is related to $\hat{\mathbf{B}}$ by a power law relationship. Hence the hysteresis energy loss in unit material volume may be expressed as

$$W_h = k_h \hat{\mathbf{B}}^N \, \text{J/m}^3 \qquad (4.10)$$

where k_h is a hysteresis loss constant, and N is called the *Steinmetz coefficient*. For materials commonly employed in electrical machinery, k_h typically lies between 150 and 400 in SI units, and N is between 1.5 and 2.5. Since the frequency f determines the rate at which the hysteresis loop is traversed, the hysteresis power loss in unit material volume follows as

$$P_h = k_h f \hat{\mathbf{B}}^N \, \text{W/m}^3 \qquad (4.11)$$

Factors influencing the magnitude of the eddy current loss may be determined by considering a block of conducting material experiencing a time-varying magnetic flux, as indicated in Figure 4.6. Figure 4.6(a) shows a general view of the block of width w, thickness t, and depth in the direction of the flux path of l. It is assumed that the flux is

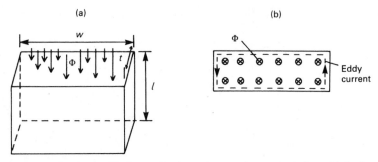

Figure 4.6 Conducting block experiencing a time-varying magnetic flux   (a) basic arrangement   (b) section showing resulting eddy-current path

uniformly distributed over the area and that t is small in comparison to w. If the flux is increasing in the direction shown, then from Faraday's and Lenz' law, a circulatory e.m.f. and current must be set up in the block tending to drive flux in opposition to the existing flux. This resulting eddy current is shown in the section of Figure 4.6(b). At power frequencies, the flux set up by the eddy currents may be neglected in comparison to the existing flux. With peak flux and flux density in the block $\hat{\Phi}$ and $\hat{B}$ respectively, the e.m.f. induced in the block, E, assuming sinusoidal excitation follows proportionally from Faraday's law as

$$E \propto \hat{\Phi} f \propto \hat{B} w t f \qquad (4.12)$$

Since the resulting eddy current flows in a circulatory path in the plane normal to the flux, then the path resistance, depending on material resistivity ρ, path length (w with $w \gg t$) and path area (tl) may be expressed proportionally as

$$R \propto \rho w / (tl) \qquad (4.13)$$

Since the power loss is E^2/R, the proportional expression for the eddy current loss P_e follows as

$$P_e \propto \hat{B}^2 f^2 w l t^3 / \rho \qquad (4.14)$$

However, the volume of the block is wlt, and so the eddy current loss in unit material volume may finally be expressed as

$$P_e = k_e \hat{B}^2 f^2 t^2 / \rho \ \text{W/m}^3 \qquad (4.15)$$

where k_e is an eddy current loss constant. A more rigorous analysis shows the theoretical value of k_e to be $\pi^2/6$ or 1.64 in SI units for a non-magnetic material under sinusoidal conditions. Nonlinearity affects the situation for magnetic materials, and values encountered here in practice range from 1.4 to 3 in SI units.

In a.c. magnetic circuits containing iron, supply frequency is not normally readily variable. Also, peak flux density is arranged to be as high as possible to make best economic use of materials, within limits set by magnetic saturation. It follows from Equation (4.11) that the only way to limit hysteresis loss in the iron is by the selection of a material with a small hysteresis loop. Such materials have in general little resistance to alternating magnetisation and are termed magnetically 'soft'. When considering eddy current loss in iron circuits, it is evident from Equation (4.15) that there are several steps which can be taken to reduce its magnitude. The material thickness t may be reduced by subdividing the solid block shown in Figure 4.6(a) into a stack of thin sheets, called *laminations*. Since P_e varies as t^2, considerable reduction in loss can be obtained using this technique. For this reason, the iron circuits of a.c. electrical machinery are almost always constructed using laminations. It is, incidentally, easier to manufacture complex magnetic circuits from thin sheets rather than solid blocks, and so the iron sections of d.c. machines are also commonly manufactured using the same technique.

An alternative way of breaking up the eddy current paths to limit eddy current loss is to subdivide the iron into small particles. The particles are held together by an inert binding material to form a *powder core* or *dust core*. These are effective at reducing loss, but suffer the disadvantage that the flux path is impaired, resulting in an overall loss of permeability.

Eddy current loss in iron may also be reduced by material treatment to increase its resistivity. The addition of controlled quantities of silicon to steel has this effect. Up to about 4% may be added, giving an increase in resistivity of approximately four times over that of relatively pure iron. The silicon reduces the machineability of the laminations, however, making them hard and brittle, and this prohibits further silicon addition.

The above techniques of iron loss reduction may be applied successfully to iron for operating frequencies up to about 400 Hz. Above this, however, eddy current loss becomes excessive, and a different class of materials must be used for devices used in radio circuits operating typically at several kHz.

These are (magnetically) *soft ferrites* which have resistivities in the semiconductor class, typically 10^6 higher than pure iron. Hence, they limit eddy current loss to manageable proportions. Their disadvantage, compared to iron, is low saturation flux density (typically 0.3 T) but they nevertheless remain the only materials generally available for the construction of magnetic circuits at these higher frequencies.

A final comment must be made concerning the use of formulae to calculate iron loss. Equations 4.11 and 4.15 apply reasonably well to samples of material under laboratory conditions. However, in a practical device, material properties may be modified by cutting and stamping operations on laminations, and small airgaps may inevitably be introduced in the assembly process. In addition, flux density varies due to leakage effects, and the insulation between laminations may be impaired due to imperfections in the insulating material used (if any), and burring at the edges of the sheets.

For these reasons, iron losses are higher than initially predicted, and a factor is normally necessary to relate iron loss formulae derived from manufacturers' characteristics to the losses occurring in real machines. The factor is obviously dependent on machine type and manufacturing technique. The discrepancy is often large, and a doubling of the iron loss over the laboratory-measured value is not atypical. More commonly, bulk-machine manufacturers dispense with formulae altogether, and use curves relating specific iron loss to peak flux density, based on experience gained from test results on previous machines of their own manufacture.

4.6 Permanent-magnet circuits

In the previous section, 'soft' magnetic materials with small hysteresis loops and little resistance to magnetisation reversal were mentioned. 'Hard' magnetic materials have the converse characteristics, and permanent magnets are the ultimate examples of these. Permanent magnets require very high magnetic field strengths to achieve magnetic saturation, and possess the ability to retain their magnetisation in the presence of demagnetising fields. They have correspondingly large hysteresis loops.

Basic permanent-magnet action may be explained by reference to Figure 4.7. Figure 4.7(a) represents a closed magnetic circuit of magnetically soft iron with one section replaced by a piece of permanent-magnet material. The permanent magnet is initially unmagnetised. In the simple analysis that follows, any leakage or fringing effects are neglected. A direct current is first applied to the coil enclosing the soft iron, of sufficient magnitude to magnetically saturate the permanent magnet. (As a practical point, it should be noted that the currents required to achieve this are normally very large, and they are usually produced by pulsed d.c. supplies). On the **B–H** characteristic of the permanent magnet shown in Figure 4.7(b), the operating point then moves from O to A. If the current is reduced again to zero, then owing

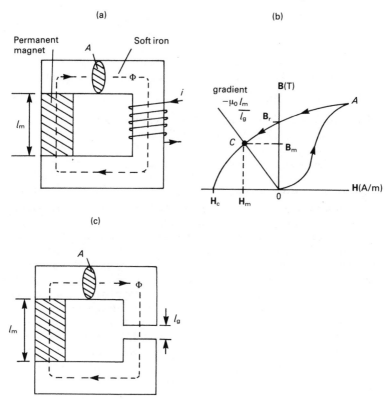

Figure 4.7 Basic operation of permanent-magnet circuit   (a) initial magnetisation
(b) permanent-magnet characteristic   (c) magnetised magnet with coil removed
and airgap inserted

to hysteresis, the operating point moves from A to $\mathbf{B_r}$ on the **B** axis. $\mathbf{B_r}$
is termed the *Remanence* of the permanent magnet. If the coil is now
removed from the soft iron, a flux density $\mathbf{B_r}$ remains in the idealised
circuit.

Insertion of an airgap of length l_g in the circuit gives it a defined
reluctance, and changes the flux level. The circuit produced is shown
in Figure 4.7(c). Application of the magnet circuit laws of Section 4.2
allows the new operating point to be calculated. Since, in the absence
of the coil, there is no explicit m.m.f. source in the magnetic circuit,
the m.m.f. equation reduces to:

$$\mathbf{H_m}l_m + \mathbf{H_g}l_g = 0 \tag{4.16}$$

where l_m is the length of the permanent magnet. But with uniform circuit area, the flux density is constant throughout the circuit and hence:

$$\mathbf{B_g} = \mu_o \mathbf{H_g} = \mathbf{B_m} \tag{4.17}$$

Substitution for $\mathbf{H_g}$ in terms of $\mathbf{B_m}$ in Equation (4.16) then gives:

$$\mathbf{B_m} = -\mu_o(l_m/l_g)\mathbf{H_m} \tag{4.18}$$

The intersection of this straight line of negative gradient with the **B–H** curve in the second quadrant gives the final operating point C, as indicated in Figure 4.7(b).

The effectiveness of a permanent-magnet material is determined by $\mathbf{B_r}$, and the intercept with the $-\mathbf{H}$ axis, termed the *coercive force* $\mathbf{H_c}$ (see Figure 4.7(b)). In general, high $\mathbf{B_r}$ values result in high circuit flux densities, and high $\mathbf{H_c}$ values give high resistance to demagnetisation. As an indication of the values likely to be encountered in practice, some typical permanent-magnet demagnetisation characteristics are shown in Figure 4.8. *Alcomax III* is a typical example of an aluminium-nickel-cobalt (Al-Ni-Co) or *metallic* magnet. These were discovered in the 1930s and reached peak production in the 1950s. They produce high magnetic fields but are prone to demagnetisation. The (magnetically) *hard-ferrite, ceramic,* or *oxide* magnets were first developed in the 1950s. They are generally cheaper than Al-Ni-Co magnets, have greater resistance to demagnetisation, but produce far inferior circuit flux densities. During the 1960s, compounds of the *rare-earth* elements with cobalt were first developed, and new ones

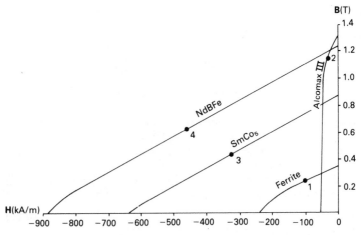

Figure 4.8 Demagnetisation characteristics for typical permanent-magnet materials

are still under development. The most widely available magnet material of this type is *samarium cobalt* ($SmCo_5$). The term 'rare' is misleading, since the cobalt is more difficult to obtain than rare-earth elements, but the rare-earth elements are expensive to refine. Rare-earth magnets have respectable remanence values but exceptional coercive force values, hence having remarkable resistance to demagnetisation. Their predictable disadvantage is high cost. Recently, a new material, *neodymium-boron-iron* (NdBFe) has appeared, which shows great promise. It has magnetic properties superior to $SmCo_5$, but is potentially significantly cheaper (see Example 4.4). Current indications are that the cost of this new material will fall as quantity production increases, and it may well be a material which will find widespread use in the near future.

In common with other magnetic circuits, permanent-magnet circuits are prone to fringing and leakage effects, and many ingenious formulae have been devised to attempt to allow for this. For details of these, and other aspects of permanent magnets, the interested reader is referred to the book by McCaig (1979).

4.7 Bibliography

McCAIG, M., *Permanent Magnets and their Applications*, Pentech Press, London (1979)

British Steel Corporation Electrical Sheet Works, *Non-oriented Electrical Steels* (1974)

RAHMAN, M. A. and SLEMON, G. R., 'Promising applications of neodymium boron iron magnets in electrical machines', *IEEE Transactions on Magnetics*, (1985) MAG–21, (5), 1712–1716

WORKED EXAMPLES

Example 4.1: CCORE: analysis of C-core with coil excitation

The effect of leakage factor on magnetic circuit performance may be considerable. Figure 4.9(a) illustrates a C-core of square cross-section constructed of mild steel. The airgap length is assumed variable, and other dimensions are given in the diagram. Various empirical formulae for leakage factor have been proposed, and one applicable to the C-core, assuming that the coil covers the whole of the bottom limb is, for the given dimensions

$$k_l = 1 + 0.1 l_g (6.8(1 - 0.1 l_g) + 9.72) \tag{4.19}$$

where the airgap length l_g is in mm. Equation (4.19), derived with simplifications from Equation (6.40) of McCaig (1979), is shown graphically in Figure 4.9(b).

(a)

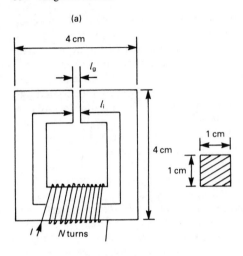

(b)

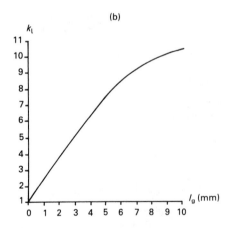

Figure 4.9 Details for Example 4.1   (a) magnetic circuit dimensions   (b) leakage factor variation with airgap

Develop a program to derive flux and flux density in the circuit from the m.m.f. applied to the coil. The **B–H** relationship for mild steel may be taken from Figure 4.2.

```
10 REM      CCORE - C-core with coil excitation
20 DATA 0,0, 0.2,100, 0.4,150, 0.6,200, 0.8,250, 1.0,400
30 DATA 1.1,500, 1.2,600, 1.3,800, 1.4,1300, 1.5,2000
40 DATA 1.55,2500, 1.6,3400, 1.65,4500, 1.7,6000
45 DATA 1.75,7500, 1.8,10000
50 DIM I1(16,1)
60 FOR J = 0 TO 16
70   FOR K = 0 TO 1
80     READ I1(J,K)
90     NEXT K
100   NEXT J
110 M0 = 4*3.14159E-7
120 A = 1E-4
130 PRINT
140 PRINT
150 PRINT "Gapped C-core with variable airgap and coil"
160 PRINT "------------------------------------------------"
170 PRINT
180 PRINT "( Core square with outside dims. 4cm x 4cm"
190 PRINT "  square cross-section 1cm x 1cm "
200 PRINT "  core material - mild steel )"
210 PRINT
220 PRINT "Airgap length (in mm)";
230 INPUT G
240 PRINT
250 PRINT "Number of turns in winding";
260 INPUT N
270 PRINT
280 PRINT "Winding current (A)";
290 INPUT I
300 PRINT
310 PRINT
320 K1 = 1.0 + 0.1*G*(6.8*(1.0 - 0.1*G) + 9.72)
330 L1 = 12E-2 - G*1E-3
340 B1 = K1*M0*N*I/(G*1E-3)
350 H1 = N*I/L1
360 IF I1(16,0) < B1 - (B1/H1)*I1(16,1) THEN GOTO 630
370 FOR J = 0 TO 16
380   IF I1(J,0) > B1 - (B1/H1)*I1(J,1) THEN GOTO 400
390   NEXT J
400 X1 = I1(J,1) - I1(J-1,1)
410 Y1 = I1(J,0) - I1(J-1,0)
420 D = Y1 + (B1/H1)*X1
430 B2 = (B1/H1)*(I1(J-1,0)*I1(J,1) - I1(J-1,1)*I1(J,0))
435 B2 = (B1*Y1 + B2)/D
440 H2 = (I1(J-1,1)*Y1 + (B1 - I1(J-1,0))*X1)/D
450 B3 = B2/K1
460 F2 = B2*A
470 F3 = B3*A
480 PRINT "Leakage factor = "; K1
490 PRINT
500 PRINT "Flux through iron inside coil = "; F2; " Wb"
510 PRINT
520 PRINT "Flux density in iron inside coil ="; B2; " T"
530 PRINT
540 PRINT "Airgap flux = "; F3; " Wb"
```

```
550 PRINT
560 PRINT "Airgap flux density = ";B3; " T"
570 PRINT
580 PRINT
590 PRINT "Another run (Y/N) ";
600 INPUT Q$
610 IF Q$ = "Y" THEN GOTO 130
620 STOP
630 PRINT "Load line does not intersect B-H curve"
640 PRINT "Input new data with reduced coil excitation"
650 GOTO 130
660 STOP
670 END
```

```
>RUN

Gapped C-core with variable airgap and coil
----------------------------------------------------

( Core square with outside dims. 4cm x 4cm
   square cross-section 1cm x 1cm
   core material - mild steel )

Airgap length (in mm)?1

Number of turns in winding?500

Winding current (A)?2.5


Leakage factor = 2.58

Flux through iron inside coil = 1.7E-4 Wb

Flux density in iron inside coil =1.7 T

Airgap flux = 6.59E-5 Wb

Airgap flux density = 0.659 T


Another run (Y/N) ?N

STOP at line 620
```

Program notes
(1) The **B–H** data for mild steel is first stored for 17 positions in the two-dimensional array I1. Pairs of values are read from DATA statements, and held in the order **B** then **H**, such that I1(0,0) holds the **B** value at the origin (zero), and I1(16,1) holds the maximum **H** value (10000).
(2) Equation (4.19) is programmed at line 320.
(3) The program implements the graphical technique of Figure 4.3. l_i is calculated (in m) at line 330, and the **B** and **H**-axis intercepts, assigned to B1 and H1 respectively, are calculated at lines 340 and 350. The

nub of the program is the determination of the intersection of the **B–H** characteristic and the straight line through the intercepts. This is performed over lines 370 and 440. Elements of array I1 are first examined in ascending order to determine the first **B** value above the straight line. Having thus established the points from the **B–H** table either side of the line, a process of *linear interpolation* is then used to calculate the approximate intersection point. This simply replaces the curve joining the two points by a straight line. X1 and Y1 are assigned to the differences in **H** and **B** values between the two points, and the intersection is then obtained from simple coordinate geometry (lines 420 to 440). B2 and H2 are then B_i and H_i, the operating conditions of the iron inside the coil.

(4) Data validation is included (line 360) to ensure that the coil m.m.f. is not so great that the straight line joining B1 and H1 is beyond the end of the stored **B–H** characteristic.

(5) Note how rapidly k_1 increases with airgap length in a circuit of this type.

Example 4.2: ROTORB: airgap flux density set up in basic rotary machine

Figure 4.10(a) represents a simplified rotary machine in which flux is set up in an airgap by a current-carrying coil on a rotor. Taking the dimensions from the diagram, and assuming the machine to be of mild steel, calculate the airgap flux density for values of rotor coil m.m.f. entered as input data. Magnetic leakage and fringing may be neglected.

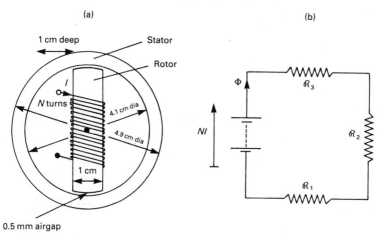

Figure 4.10 Details for Example 4.2   (a) dimensions of machine   (b) magnetic equivalent circuit

```
10 REM        ROTORB - Flux density set up by rotor coil
20 REM                 with magnetically non-linear iron
30 DIM R(3)
40 DIM K1(3)
50 DIM B(3)
60 DIM H(3)
70 DIM M(3)
80 DATA 0,0, 0.2,100, 0.4,150, 0.6,200, 0.8,250, 1.0,400
90 DATA 1.1,500, 1.2,600, 1.3,800, 1.4,1300, 1.5,2000
100 DATA 1.55,2500, 1.6,3400, 1.65,4500
110 DATA 1.7,6000, 1.75,7500, 1.8,10000, 2.0,30000
120 DIM I1(17,1)
130 FOR J = 0 TO 17
140    FOR K = 0 TO 1
150       READ I1(J,K)
160       NEXT K
170    NEXT J
180 P1 = 3.14159265
190 M0 = 4*P1*1E-7
200 E0 = 0.001
210 A = 0.1
220 DATA 4.5E-2, .5E-3, 1.1E-2, 1E-2, 4E-2, 1E-2, 4E-3
230 READ D1, L1, L2, L3, L4, L5, L6
240 PRINT
250 PRINT
260 PRINT "Flux density set up in basic rotary device"
270 PRINT "--------------------------------------------"
280 PRINT
290 PRINT "Dimensions (in m):- mean stator dia = "; D1
300 PRINT
310 PRINT "airgap length = "; L1; " pole length = "; L2
320 PRINT
330 PRINT "axial length = "; L3; " rotor length = "; L4
340 PRINT
350 PRINT "rotor width = "; L5; " stator width = "; L6
360 PRINT
370 PRINT "(magnetic material assumed to be mild steel)"
380 PRINT
390 PRINT "Enter rotor coil m.m.f (AT) ";
400 INPUT F
410 R(0) = 2*L1/(M0*L2*L3)
420 K1(1) = L4/(L3*L5)
430 K1(2) = P1*D1/(2*L3*L6)
440 FOR K = 1 TO 2
450    M(K) = 2E-4
460    NEXT K
470 N = 0
480 M1 = M(1)
490 M2 = M(2)
500 FOR K = 1 TO 2
510    R(K) = K1(K)/M(K)
520    NEXT K
530 P0 = F/(R(0) + R(1) + R(2))
540 B(1) = P0/(L3*L5)
550 B(2) = P0/(2*L3*L6)
560 FOR K = 1 TO 2
570    FOR J = 0 TO 17
580       IF B(K)>2.0 THEN B(K) = 1.999999
590       IF I1(J,0) > B(K) THEN GOTO 610
```

```
600      NEXT J
610      X1 = I1(J,1) - I1(J-1,1)
620      Y1 = I1(J,0) - I1(J-1,0)
630      H(K) = I1(J-1,1) + (B(K) - I1(J-1,0))*X1/Y1
640      M(K) = A*B(K)/H(K) + (1-A)*M(K)
650      NEXT K
660  N = N+1
670  E1 = ABS((M1 - M(1))/M1)
680  E2 = ABS((M2 - M(2))/M2)
690  IF E1>E0 THEN GOTO 480
700  IF E2>E0 THEN GOTO 480
710  IF B(2)>=1.999999 THEN GOTO 840
720  B0 = P0/(L2*L3)
730  PRINT
740  PRINT
750  PRINT "Airgap flux density = "; B0; " T"
760  PRINT
770  PRINT "( Convergence to "; E0 " achieved in "; N;
775  PRINT " iterations )"
780  PRINT
790  PRINT
800  PRINT "Another run (Y/N) ";
810  INPUT Q$
820  IF Q$ = "Y" THEN GOTO 240
830  STOP
840  PRINT
850  PRINT
860  PRINT "B-H data exceeded due to high saturation"
870  PRINT "Please try again with reduced excitation"
880  GOTO 380
890  END
```

```
>RUN

Flux density set up in basic rotary device
-------------------------------------------

Dimensions (in m):- mean stator dia = 4.5E-2

airgap length = 5E-4 pole length = 1.1E-2

axial length = 1E-2 rotor length = 4E-2

rotor width = 1E-2 stator width = 4E-3

(magnetic material assumed to be mild steel)

Enter rotor coil m.m.f (AT) ?1500


Airgap flux density = 1.18 T.

( Convergence to 1E-3 achieved in 43 iterations )


Another run (Y/N) ?N

STOP at line 830
```

Program notes

(1) The magnetic equivalent circuit of the device is shown in Figure 4.10(b). $\mathscr{R}_1$ represents the reluctance of the two airgaps in series, $\mathscr{R}_2$ is the rotor reluctance, and $\mathscr{R}_3$ the stator reluctance. The formulae using the notation of the program are

$$\mathscr{R}_1 = 2.L1/(M0.L2.L3) \tag{4.20}$$

$$\mathscr{R}_2 = L4/(M(1).L3.L5) \tag{4.21}$$

$$\mathscr{R}_3 = \pi.D1/(2.M(2).L3.L6) \tag{4.22}$$

where M(1) and M(2) are the permeabilities of rotor iron and stator iron respectively. Note the parallel flux paths of equal reluctance on the stator.

The circuit could be simply solved for flux and hence airgap flux density if M(1) and M(2) were constants. However, for the magnetically non-linear mild steel, they are variables, their values dependent on the appropriate flux densities. This difficulty may be overcome by use of an iterative process. Initial values are assumed for M(1) and M(2) (in this case 2×10^{-4}) and the circuit solved for flux. The resultant flux densities imply new values of M(1) and M(2) which may be substituted and the process repeated. In many situations, this technique results in the variables converging to steady values, and the answer may be assumed correct when the change occurring is less than some predefined error value. In this case, however, the process is highly unstable since small changes in flux density can result in rapid changes in implied permeability. Hence, no steady values would be obtained. This problem is solved by limiting the change demanded at each iteration. An under-relaxation factor A is introduced (< 1) such that the change in M(1) and M(2) is reduced to A of what it would otherwise be. For a variable V, the new value employed is therefore related to the old by the expression

$$V_{employed} = A.V_{new} + (1 - A).V_{old} \tag{4.23}$$

Evidently at the end of the process, when V_{new} and V_{old} are converging, (4.23) in the limit has no effect, which is the desired situation.

(2) Three-element arrays R, K1, B, H, and M are used to hold reluctance, geometric constants, **B**, **H**, and permeability values respectively for the three sections of the magnetic circuit. The two-dimensional array I1 is employed as in CCORE, with the exception that one extra pair of **B–H** values is stored to aid the iteration process.

(3) E0 is an error limit, such that the iterative process is considered complete when the per unit change in both M(1) and M(2) from iteration to iteration is less than E0. E0 is preset to 0.001 at line 200.

(4) The iterative process covers lines 480–700. The loop includes the variable N to count the number of iterations required. The magnetic

equivalent circuit is actually solved at line 530 to give the flux P0. M1 and M2 store the current values of steel permeability which are updated at line 640, using the principle of Equation (4.23). Linear interpolation is again employed to calculate H and then M values from the B values obtained at lines 540 and 550. The coordinate geometry is simpler in this case, and spans lines 610–630. The error tests cover lines 670–700.

(5) During the iteration process, the implied flux density may exceed the maximum value stored (2.0 T). In this event, the value is limited to stay within the permissible range (line 580). If, however, the final stator flux density remains limited to 2.0 T, then insufficient data points are available for the B–H characteristic, and the user is requested to reduce the coil excitation.

(6) The best value of A is found by experiment. The value of 0.1 employed guarantees stability and gives convergence in 40 to 70 iterations, with a general increase in iteration number with m.m.f. and hence magnetic saturation in the device.

Example 4.3: FELOSS: iron loss variation with material, f and $\hat{B}$

Equations (4.11) and (4.15) may be combined to give an expression for total specific iron loss of the form

$$P = k_e(\hat{B}ft)^2 + k_h f \hat{B}^N \text{ (W/kg)} \tag{4.24}$$

British Steel Corporation (1974) contains curves of specific iron loss for various lamination materials which may be represented approximately by the following values for constants in the above equation.

Material	t (mm)	k_e	k_h	N
Newcor 1000	0.65	1473	.04469	2.2142
Newcor 800	0.65	1261	.03382	2.4365
Losil 800	0.65	1065	.0424	1.9063
Losil 630	0.65	1031	.03367	1.8769
Losil 500	0.5	768	.0361	1.752
Losil 400	0.5	658	.02606	1.8548
Transil 335	0.35	526	.02849	1.8889

Write a program to tabulate the specific iron loss for any of the materials and any frequency for a range of peak flux density values. $\hat{B}$ should be varied from 0.2 T to 2 T, and the program should include a multiplying factor of 2 as an allowance for increase due to manufacturing processes etc.

```
 10 REM       FELOSS - Effect of frequency and peak
 20 REM       flux density on specific iron loss
 25 REM       for various electrical steels
 30 DIM I$(7), D(7,4)
 40 K1 = 2.0
 50 DATA "Newcor 1000", "Newcor 800", "Losil 800"
 60 DATA "Losil 630", "Losil 500", "Losil 400"
 65 DATA "Transil 335"
 70 FOR J = 1 TO 7
 80    READ I$(J)
 90    NEXT J
100 DATA .65,    1473,   .04469,   2.2142
110 DATA .65,    1261,   .03382,   2.4365
120 DATA .65,    1065,   .0424,    1.9063
130 DATA .65,    1031,   .03367,   1.8769
140 DATA .50,     768,   .0361,    1.752
150 DATA .50,     658,   .02606,   1.8548
160 DATA .35,     526,   .02849,   1.8889
170 FOR J = 1 TO 7
180    FOR K = 1 TO 4
190       READ D(J,K)
200       NEXT K
210    NEXT J
220 PRINT
230 PRINT
240 PRINT "Iron loss with adjustment factor ";K1
250 PRINT "----------------------------------------"
260 PRINT
270 PRINT "Operating frequency (Hz) ";
280 INPUT F
290 PRINT
300 PRINT "The lamination materials are identified"
310 PRINT "by numbers as follows:"
320 PRINT
330 FOR J = 1 TO 7
340    PRINT "For "; I$(J); TAB(17); " the number is "; J
350    NEXT J
360 PRINT
370 PRINT "Code number of the material of interest ";
380 INPUT N
390 T = D(N,1)/1000
400 PRINT
410 PRINT
420 PRINT "Iron-loss data for "; I$(N); " - "; D(N,1);
425 PRINT "mm at "; F; "Hz"
430 PRINT "---------------------------------------------"
440 PRINT
450 PRINT "Peak flux density (T)        Iron loss (W/kg)"
460 PRINT TAB(25); "( "; K1; " x ideal value )"
470 PRINT
480 FOR B = 0.2 TO 2.1 STEP 0.2
490    P = K1*(D(N,2)*(F*T*B)^2 + D(N,3)*F*B^D(N,4))
500    PRINT TAB(8); B; TAB(28); P
510    NEXT B
520 PRINT
530 PRINT "Another material (Y/N) ";
540 INPUT Q$
550 IF Q$ = "Y" THEN GOTO 360
560 PRINT
570 PRINT "Another frequency (Y/N) ";
580 INPUT Q$
590 IF Q$ = "Y" THEN GOTO 260
```

```
600  STOP
610  END


>RUN


Iron loss with adjustment factor 2
----------------------------------------

Operating frequency (Hz) ?50

The lamination materials are identified
by numbers as follows:

For Newcor 1000    the number is 1
For Newcor 800     the number is 2
For Losil 800      the number is 3
For Losil 630      the number is 4
For Losil 500      the number is 5
For Losil 400      the number is 6
For Transil 335    the number is 7

Code number of the material of interest ?1

Iron-loss data for Newcor 1000 - 0.65mm at 50Hz
----------------------------------------------------

Peak flux density (T)        Iron loss (W/kg)
                             ( 2 x ideal value )

        0.2                      0.251
        0.4                      1.09
        0.6                      2.56
        0.8                      4.72
        1                        7.58
        1.2                      11.2
        1.4                      15.5
        1.6                      20.6
        1.8                      26.5
        2                        33.2

Another material (Y/N) ?Y

Code number of the material of interest ?7

Iron-loss data for Transil 335 - 0.35mm at 50Hz
----------------------------------------------------

Peak flux density (T)        Iron loss (W/kg)
                             ( 2 x ideal value )

        0.2                      0.149
        0.4                      0.556
        0.6                      1.2
        0.8                      2.08
        1                        3.17
        1.2                      4.48
        1.4                      6.01
        1.6                      7.75
        1.8                      9.69
        2                        11.8
```

```
Another material (Y/N) ?N

Another frequency (Y/N) ?Y

Operating frequency (Hz) ?400

The lamination materials are identified
by numbers as follows:

For Newcor 1000    the number is 1
For Newcor 800     the number is 2
For Losil 800      the number is 3
For Losil 630      the number is 4
For Losil 500      the number is 5
For Losil 400      the number is 6
For Transil 335    the number is 7

Code number of the material of interest ?5


Iron-loss data for Losil 500 - 0.5mm at 400Hz
---------------------------------------------------------

Peak flux density (T)          Iron loss (W/kg)
                              ( 2 x ideal value )

        0.2                        4.18
        0.4                        15.6
        0.6                        33.9
        0.8                        58.9
        1                          90.3
        1.2                        128
        1.4                        172
        1.6                        223
        1.8                        280
        2                          343

Another material (Y/N) ?N

Another frequency (Y/N) ?N

STOP at line 600
```

Program notes
(1) A two-dimensional array D holds the data from the above table. The material names are held in the string array I$.
(2) Since the process of entering and comparing long strings is prone to error, materials are selected by a numeric code from 1 to 7. Lines 300–350 cover the display of this code for the user.
(3) The run first compares loss in Newcor 1000 and Transil 335 at 50 Hz. Note the reduction to almost 1/3 of the loss obtained with the thinner lamination and silicon additive of the latter. The order of magnitude increase in iron loss shown with the 400 Hz run for Losil 500 indicates the high frequency limitations of these materials.

Example 4.4: MAGNETS: comparison of permanent magnet materials

A permanent magnet is required to set up a prescribed flux density in an airgap of given physical dimensions. The leakage factor may be assumed known, and independent of the shape and size of magnet required. Magnet materials available are ferrite, Alcomax III, samarium cobalt ($SmCo_5$), and neodymium-boron-iron (NdBFe). For each material, the magnetic circuit is tailored to enable the permanent magnet to operate at a suitable fixed point ($\mathbf{B_m}, \mathbf{H_m}$) on its demagnetisation characteristic (see points 1 to 4 on Figure 4.8). Relative specific cost, derived from Rahman and Slemon (1985), and specific gravity complete the data in the table below.

Material	$\mathbf{B_m}$ (T)	$-\mathbf{H_m}$ (kA/m)	S.G.	Rel. cost/kg
Ferrite	0.24	100	5.0	1
Alcomax III	1.15	33	7.3	7
$SmCo_5$	0.44	325	8.3	20
NdBFe	0.63	465	7.4	11

Write a program to calculate magnet dimensions, weight and relative cost for each of the permanent magnet materials.

```
10 REM       MAGNETS - Comparitive performance of typical
20 REM                 permanent magnets in setting up
25 REM                 flux in an airgap
30 DIM I$(4), D(4,5), L(4), A1(4), W(4), V(4)
40 P1 = 3.14159265
50 MO = 4*P1*1E-7
60 DATA "Ferrite", "Alcomax III", "SmCo5", "NdBFe"
70 FOR J = 1 TO 4
80    READ I$(J)
90    NEXT J
100 DATA 0.24,    100E3,    5.0,    1
110 DATA 1.15,     33E3,    7.3,    7
120 DATA 0.44,    325E3,    8.3,    20
130 DATA 0.63,    465E3,    7.4,    11
140 FOR J = 1 TO 4
150    FOR K = 1 TO 4
160       READ D(J,K)
170       NEXT K
180    NEXT J
190 PRINT
200 PRINT
210 PRINT "Magnets setting up airgap flux"
220 PRINT "-----------------------------------"
230 PRINT
```

```
240 PRINT "Airgap length (mm) ";
250 INPUT L1
260 PRINT
270 PRINT "c.s.a. of airgap (cm^2) ";
280 INPUT A
290 PRINT
300 PRINT "Leakage factor ";
310 INPUT K1
320 PRINT
330 PRINT "Flux density required in airgap (T) ";
340 INPUT B2
350 L1 = L1*1E-3
360 A = A*1E-4
370 PRINT
380 PRINT TAB(11); "Ferrite      Alcomax III    SmCo5";
385 PRINT "            NdBFe"
390 PRINT TAB(11); "-------      -----------    -----";
395 PRINT "            -----"
400 PRINT
410 PRINT "Mag L (cm)";
420 FOR J = 1 TO 4
430    L(J) = B2*L1/(M0*D(J,2))
440    PRINT TAB(12*J); L(J)*100;
450    NEXT J
460 PRINT
470 PRINT
480 PRINT "Mag A(cm^2)";
490 FOR J = 1 TO 4
500    A1(J) = K1*A*B2/D(J,1)
510    PRINT TAB(12*J); A1(J)*1E4;
520    NEXT J
530 PRINT
540 PRINT
550 PRINT "Mag V(cm^3)";
560 FOR J = 1 TO 4
570    V(J) = L(J)*A1(J)
580    PRINT TAB(12*J); V(J)*1E6;
590    NEXT J
600 PRINT
610 PRINT
620 PRINT "Mag wt (gm)";
630 FOR J = 1 TO 4
640    W(J) = V(J)*D(J,3)
650    PRINT TAB(12*J); W(J)*1E6;
660    NEXT J
670 PRINT
680 PRINT
690 PRINT "Rel cost ";
700 FOR J = 1 TO 4
710    PRINT TAB(12*J); W(J)*D(J,4)/(W(1)*D(1,4));
720    NEXT J
730 STOP
740 END
```

```
>RUN

Magnets setting up airgap flux
---------------------------------

Airgap length (mm) ?.5

c.s.a. of airgap (cm^2) ?1

Leakage factor ?1.5

Flux density required in airgap (T) ?.5

                Ferrite     Alcomax III    SmCo5       NdBFe
                -------     -----------    -----       -----

Mag L (cm)   0.1989        0.6029        6.121E-2    4.278E-2

Mag A(cm^2) 3.125          0.6522        1.705       1.19

Mag V(cm^3) 0.6217         0.3932        0.1043      5.093E-2

Mag wt (gm) 3.108          2.87          0.866       0.3769

Rel cost     1             6.463         5.572       1.334
STOP at line 730
```

Program notes
(1) Data for the four permanent-magnet materials is held in a similar manner to that used to store iron loss data in previous program FELOSS.
(2) With H_m known for the magnet, magnet length, held in array L, follows from Equation (4.16) and the airgap **B–H** relationship. The corresponding assignment statement is at line 430. Note in the subsequent PRINT statement, the use of multiple TAB spacing, the change of unit, and the trailing semi-colon to achieve continuous printing of magnet lengths on the same line.
(3) Magnet area (array A1) follows directly from B_m and the flux continuity equation, with the effect of leakage factor included.
(4) With magnet length inversely proportional to H_m, and area inversely proportional to B_m, the $B_m H_m$ product should be as high as possible to minimise volume of magnet material required. For a linear demagnetisation characteristic, the optimum is apparently achieved by operation in the middle. This is termed the $(BH)_{max}$ point. Although ideal in theory, $(BH)_{max}$ operation of magnets is, for a variety of reasons, not often achieved in practice. For the ferrite, point 1 on Figure 4.8 is above this optimum because the material is on the verge of permanent demagnetisation at that point (i.e. the characteristic is no longer linear). For Alcomax III, $(BH)_{max}$ working is self-

evidently quite impractical due to the shape of the characteristic. However, the virtually straight-line characteristics of SmCo$_5$ and NdBFe make optimum operation possible, and points 3 and 4 on Figure 4.8 correspond to mimimum volume conditions for these two materials.

(5) From the theory and the sample run, it should be evident that the shape of magnet is governed by its operating point $\mathbf{B_m}$, $\mathbf{H_m}$. A high $\mathbf{B_m}$ and low $\mathbf{H_m}$, as with Alcomax III, results in a long thin magnet, whereas the converse, applicable for the other materials, produces a short fat magnet.

(6) With a fixed operating point for each material, the ratio of volume and weight of permanent magnet required is independent of the input data. Hence the relative cost is always the same. The program confirms the promise of the new material NdBFe; with the data used the material used is little more than one tenth the weight of the ferrite for a cost increase of only about one third. Magnet material costs are variable, however, and depend to a considerable extent on volume of manufacture. The success of NdBFe will hence hinge on the willingness of equipment manufacturers to modify designs to accommodate it.

PROBLEMS

(4.1) Develop a program to calculate permeability ($\mu_o \mu_r$) and relative permeability for mild steel. The **B–H** data from CCORE should be used, and output should be displayed for **B**-values from 0.8 to 1.8 T in steps of 0.2 T. Note particularly the shape of the curve and the orders of magnitudes of the values obtained.

(4.2) Repeat the above problem for **H**-values of 500 A/m to 10 000 A/m in steps of 500 A/m. This will entail the use of linear interpolation.

(4.3) Develop a program to solve a magnetic circuit of the form of Figure 4.4 using an equivalent-circuit approach. Input data should include excitation m.m.f., and effective lengths, cross-sectional areas and relative permeabilities (assumed constant) of the constituent limbs and airgap. Magnetic flux leakage and fringing may be neglected, and output should be flux and flux density in the airgap.

(4.4) The converse problem may be solved in stages using conventional magnetic circuit analysis methods. With physical dimensions input as before, write a program to calculate the excitation required to produce a given airgap flux density. Core material is mild steel with the **B–H** data as used in program CCORE. With reference to Figure 4.4(a), airgap m.m.f. follows directly, and the m.m.f. for the right-hand limb (carrying Φ_2) may be obtained from the limb length and linear interpolation of the **B–H** data. The centre-limb m.m.f. is the sum of the two, from which Φ_1 may be derived via the **B–H** data again. The remainder of the analysis should be self-evident from study of Figure 4.4.

(**4.5**) The specific iron loss of a stack of laminations was measured at a constant frequency of 50 Hz with three different values of peak sinusoidal flux density. Results were as follows

$\hat{B}(T)$	0.5	1.0	1.5
$P(W/kg)$	0.965	4.25	10.133

A further test was performed with the frequency raised to 60 Hz when the loss at a $\hat{B}$ value of 1T was measured at 5.4 W/kg. Assuming the iron loss to be represented by Equation (4.24), calculate values for the constants, and hence write a program to calculate and display specific iron loss for the laminations for any value of f or $\hat{B}$.

(**4.6**) A method of calculating the operating point of a permanent magnet maintaining a flux in an airgap is described in Section 4.6. Develop a program to implement the technique, covering the four materials shown in Figure 4.8. A circuit of the form of Figure 4.7(c) is to be assumed. The soft iron may be taken as having infinite permeability, and the effects of leakage flux should be included. Input data required is magnet and airgap length, circuit cross-sectional-area, leakage factor, and a numeric code to determine which magnet material is under consideration. Suitable pairs of **B–H** values may be determined from Figure 4.8 and subsequently stored. Linear interpolation is then required to determine permanent magnet operating points.

The transformer

ESSENTIAL THEORY

5.1 Introduction

The transformer is an a.c. device used to change voltage levels in electrical power-transmission systems. High voltages reduce I^2R losses and conductor cross-sections for transmission purposes. However, for safety reasons and the requirements of reasonable conductor separations, voltages must be reduced for distribution to individual consumers. In a.c. systems, the transformer effects this change simply and efficiently. The transformer also isolates circuit sections electrically from each other, which has safety benefits. It is also used in low power circuits to transform impedance values for power-matching purposes.

In this chapter, the performance equations of a perfect or *ideal* transformer are developed. The imperfections of real transformers are then progressively introduced, and an equivalent circuit evolved to allow for these. Tests to determine values of these equivalent circuit components are described, and the performance of real transformers is discussed.

5.2 The ideal two-winding transformer

The ideal two-winding transformer is shown in Figure 5.1(a). It consists essentially of two coils linked magnetically by a ferromagnetic core. The term 'ideal' implies perfect magnetic coupling between the two coils (i.e. no leakage flux), zero resistance windings, infinite permeability iron, and no iron losses. The last two conditions result in zero supply current when the load is disconnected. By convention, the coil connected to the supply is termed the *primary*, and the coil connected to the load the *secondary*.

The most fundamental property governing transformer operation is its *turns ratio*. With an a.c. supply, the flux alternates, inducing voltages in the two coils. Then from Faraday's law:

$$E_1 = N_1 \frac{d\Phi}{dt} \text{ and } E_2 = N_2 \frac{d\Phi}{dt} \tag{5.1}$$

(a)

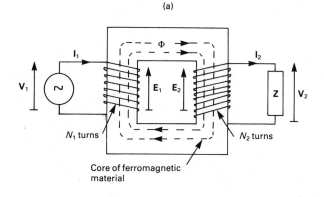

(b)

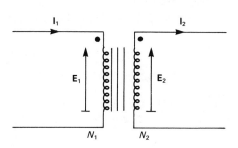

Figure 5.1 The ideal two-winding transformer   (a) physical arrangement
(b) circuit representation

But, with perfect magnetic coupling, the same flux links both coils,
and it follows that:

$$\frac{E_1}{E_2} = \frac{N_1}{N_2} \tag{5.2}$$

Hence, the voltage is transformed in the same ratio as the turns. This
is the basic operating equation of the two-winding transformer.

Under sinusoidal operating conditions, the induced e.m.f. may be
readily expressed in terms of the frequency, turns, and peak flux. The
flux may be represented as:

$$\Phi = \hat{\Phi} \sin 2\pi f t \tag{5.3}$$

where $\hat{\Phi}$ is the peak flux in the core. Then from equation (5.1):

$$E_1 = 2\pi f N_1 \hat{\Phi} \cos 2\pi f t \tag{5.4}$$

The r.m.s. induced voltage $\mathbf{E}_1$ then follows directly as

$$\mathbf{E}_1 = \sqrt{2\pi f N_1 \hat{\Phi}} = 4.44 f N_1 \hat{\Phi} \qquad (5.5)$$

and similarly for the secondary:

$$\mathbf{E}_2 = 4.44 f N_2 \hat{\Phi} \qquad (5.6)$$

In the ideal transformer, $\mathbf{V}_1 = \mathbf{E}_1$ in phasor notation, since there is no winding impedance. Hence, $\mathbf{E}_1, f$, and N_1 are fixed by the supply conditions and the particular transformer design. It follows from Equation (5.5) that $\hat{\Phi}$ must also be constant. Hence, *the ideal transformer operates at constant flux*.

If a load is connected to the ideal transformer, a current $\mathbf{I}_2$ must flow in the secondary winding, and an m.m.f. $N_2 I_2$ is available to drive flux around the core. However, no nett m.m.f. is required to set up the flux in an ideal transformer. Hence, an equal and opposite m.m.f. must be set up by the primary to cancel out the secondary m.m.f. The resulting primary current follows from the m.m.f. equation

$$N_1 I_1 = N_2 I_2 \qquad (5.7)$$

This is known as the *principle of ampère-turn balance* for a transformer.

Equations (5.2) and (5.7) determine the performance of the ideal transformer, and it is represented by the circuit shown in Figure 5.1(b).

5.3 Equivalent circuits for real transformer

A well-designed transformer is an efficient and effective device, and the ideal model developed above gives a reasonable representation of its performance. However, real transformers do have some magnetic flux leakage, windings have finite resistance, and the iron both requires a small m.m.f. to set up the flux, and suffers iron loss under a.c. conditions. For more accurate analysis, these effects must be allowed for. Each can be conveniently represented by a network component appropriately connected to the ideal transformer circuit to produce an overall equivalent circuit for a real transformer.

The effects of real iron circuits will be considered first. Some primary current is required to set up the flux for two reasons; real iron is not infinitely permeable, and some small airgaps in the magnetic circuit are normally unavoidable due to the manufacturing techniques used. Hence, these effects may be represented in a real transformer with open-circuited secondary as indicated in Figure 5.2(a). If current I_{om} flows in this case, then from the magnetic circuit laws:

(a)

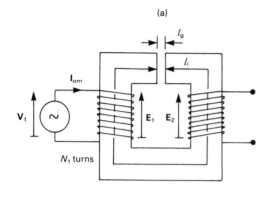

(b)

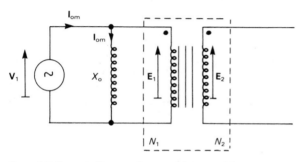

Figure 5.2 Two-winding transformer with magnetising current   (a) physical
representation   (b) equivalent circuit representation

$$\mathbf{H}_i l_i + \mathbf{H}_g l_g = N_1 I_{om} \tag{5.8}$$

The flux in the circuit depends on the supply voltage and follows from
Equation (5.5). Flux density is then determined by the core cross-
sectional area. If estimates may be made for l_i and l_g, then I_{om} may be
calculated using the technique described in Section 4.3 of the previous
chapter. Note that the problem is now an a.c. one, and that the non-
linear **B–H** relationship of the iron means that the sinusoidal Φ must
be set up by a non-sinusoidal I_{om}. However, I_{om} is small, being typically less than 5% of full-load current in a well-designed transformer.
Hence, the effect can be adequately allowed for in an equivalent circuit by a constant inductance and phasor. Since the windings of the
ideal transformer have zero impedance, and it draws no current from

the supply on no load, this inductance must be placed in parallel with the primary winding to provide a path for I_{om}. The resulting circuit is shown in Figure 5.2(b). The section within the dotted box is the ideal transformer, and the inductive reactance X_o carrying the magnetising current I_{om} is termed the *magnetising reactance*.

The second effect of real iron is the introduction of iron losses under the a.c. flux conditions. In Section 4.5 of the previous chapter, the two principal components of iron loss, hysteresis and eddy-current, were examined. If transformer operation is restricted to fixed frequency only (the normal case), then hysteresis power loss is proportional to $\hat{B}^N$ (Equation (4.11)), where N is between 1.5 and 2.5 for ferromagnetic material used in conventional transformers. Eddy current power loss is proportional to $\hat{B}^2$ (Equation (4.15)). Hence, to a reasonable approximation, the transformer iron loss is proportional to $\hat{B}^2$ or $\hat{\Phi}^2$. It then follows from the transformer voltage equation, that the iron loss on no-load is proportional to E_1^2. Hence, the iron loss may be represented in the equivalent circuit by a resistance R_o connected in parallel with the ideal transformer primary winding. The iron loss is then E_1^2/R_o.

The equivalent circuit allowing for both core magnetisation and iron loss is shown in Figure 5.3(a). The current I_{oc} flowing through R_o is the core-loss component of the no-load current I_o. Note that the characteristics of the ideal transformer (within the dotted box of Figure 5.3(a)) are still preserved; no current flows through it on no-load, with the no-load current dividing between X_o and R_o such that

$$I_o = I_{om} + I_{oc} \tag{5.9}$$

The corresponding no-load phasor diagram for the real transformer is shown in Figure 5.3(b). In practice, I_{om} is considerably greater than I_{oc}. The no-load phase angle is therefore large (typically 70° or greater), and the no-load power factor $\cos \varphi_o$ correspondingly low.

In the real transformer, imperfect magnetic coupling results in leakage flux from primary and secondary windings. This situation is illustrated in Figure 5.4(a) for a transformer connected to a load. The flux Φ linking both primary and secondary induces the e.m.f.s E_1 and E_2. However, primary current I_1 drives flux in the direction of Φ, and Φ_{1_l} represents the leakage flux set up by I_1 which links the primary but not the secondary. Conversely, secondary current I_2 drives flux in opposition to I_1, and hence in the opposite sense to Φ. Φ_{2_l} represents the secondary leakage flux which does not link the primary. The overall effect is therefore to *increase* the flux linking the primary winding, and to *reduce* that linking the secondary. The supply and load voltages are induced by the total primary and secondary fluxes respectively, and hence

(a)

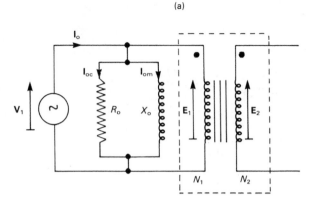

(b)

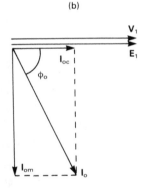

Figure 5.3 Two-winding transformer with magnetising current and iron loss
(a) equivalent circuit   (b) phasor diagram

$$V_1 = N_1 \frac{\mathrm{d}}{\mathrm{d}t}(\Phi + \Phi_{1l}) = E_1 + N_1 \frac{\mathrm{d}\Phi_{1l}}{\mathrm{d}t} \tag{5.10}$$

$$V_2 = N_2 \frac{\mathrm{d}}{\mathrm{d}t}(\Phi - \Phi_{2l}) = E_2 - N_2 \frac{\mathrm{d}\Phi_{2l}}{\mathrm{d}t} \tag{5.11}$$

But Φ_{1l} and Φ_{2l} pass largely through air, and are little affected by
saturation. They may therefore be considered to be proportional to
currents setting them up, and may be represented by constant in-
ductances such that $N_1\Phi_{1l} = L_1 I_1$ and $N_2\Phi_{2l} = L_2 I_2$. L_1 and L_2 are
termed the primary and secondary *leakage inductance* respectively.
Equations (5.10) and (5.11) may then be written in the final form

(a)

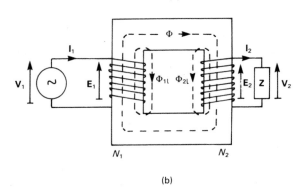

(b)

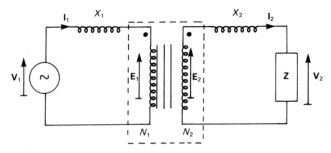

Figure 5.4 Two-winding transformer including the effects of leakage flux
(a) physical representation   (b) equivalent circuit

$$V_1 = E_1 + L_1 \frac{dI_1}{dt} \tag{5.12}$$

$$V_2 = E_2 - L_2 \frac{dI_2}{dt} \tag{5.13}$$

This is represented in equivalent circuit form in Figure 5.4(b). The components $X_1 = \omega L_1$ and $X_2 = \omega L_2$ follow as the primary and secondary *leakage reactance* respectively.

The overall effect of leakage flux is to change the value of V_2 as I_2 and Z change. It is important to note however that, having no associated resistive component, it introduces no power loss.

The final property of a real transformer to be represented is winding resistance. This has a two-fold effect on performance; the introduction of volt drops on both sides of the transformer, and the dissipation of power as copper loss. It may be readily incorporated in the equivalent circuit by the introduction of the appropriate resistances in series with the winding leakage reactances.

The complete equivalent circuit for the two-winding transformer is shown in Figure 5.5. It is sometimes referred to as an 'exact' equivalent circuit. Note that the current flowing in the primary of the ideal transformer, denoted by $\mathbf{I}_1'$, is given by the equation $\mathbf{I}_1'N_1 = \mathbf{I}_2 N_2$.

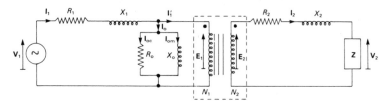

Figure 5.5 'Exact' equivalent circuit of real two-winding transformer

5.4 Referred impedances

Analysis of the complete equivalent circuit is complicated by the 'non-ideal' current $\mathbf{I}_1$ flowing through the primary series components R_1 and X_1. However, in a well-designed transformer, the voltage drop in the primary series components is only a small fraction of $\mathbf{V}_1$, even under full-load conditions. The voltage across the R_o/X_o parallel combination ($\mathbf{E}_1$) is therefore little different from $\mathbf{V}_1$, and so R_o/X_o may be reconnected directly across the supply terminals with little loss of accuracy. This results in the 'approximate' equivalent circuit shown in Figure 5.6, which has the ideal current $\mathbf{I}_1'$ flowing through the primary components R_1 and X_1. It should be appreciated that the terms 'exact' and 'approximate' are merely relative; the exact circuit is not a precise model of real transformer behaviour, but it is more accurate than the approximate circuit.

It is often convenient to transfer, or refer, series impedances from one side to the other of our equivalent circuit. This is possible in the approximate circuit which, from Figure 5.6, has equations as follows:

$$\mathbf{V}_1 = \mathbf{E}_1 + (R + jX_1)\mathbf{I}_1' \qquad (5.14)$$

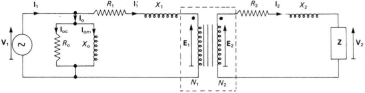

Figure 5.6 'Approximate' equivalent circuit of real two-winding transformer

$$E_2 = V_2 + (R_2 + jX_2)I_2 \tag{5.15}$$

But using the ideal transformer equations $E_2 = E_1 \ (N_2/N_1)$ and $I_2 = I_1'(N_1/N_2)$, Equation (5.15) may be re-arranged to give an expression for E_1 in terms of V_2 and I_1'. Substitution for E_1 in Equation (5.14) then gives the final form:

$$V_1 = \left\{ R_1 + \left\{ \frac{N_1}{N_2} \right\}^2 R_2 + j(X_1 + \left\{ \frac{N_1}{N_2} \right\}^2 X_2) \right\} I_1' + \left\{ \frac{N_1}{N_2} \right\} V_2 \tag{5.16}$$

This equation results in the modified approximate equivalent circuit shown in Figure 5.7. In this circuit, both the secondary equiva-

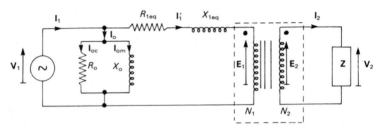

Figure 5.7 Approximate equivalent circuit with series components referred to primary side

lent circuit components have been replaced by equivalent forms on the primary side. From Equation (5.16):

$$R_{1eq} = R_1 + \left\{ \frac{N_1}{N^2} \right\}^2 R_2 \tag{5.17}$$

and

$$X_{1eq} = X_1 + \left\{ \frac{N_1}{N^2} \right\}^2 X_2 \tag{5.18}$$

where R_{1eq} and X_{1eq} are termed respectively the *equivalent resistance referred to the primary* and the *equivalent leakage reactance referred to the primary*. It is customary to indicate referred values by a prime (') so that $R_2' = R_2 \left(\dfrac{N_1}{N_2} \right)^2$ is the secondary winding resistance referred to the primary side.

Evidently, the primary series components may be similarly referred to the secondary side to give the corresponding approximate equivalent circuit shown in Figure 5.8. Here:

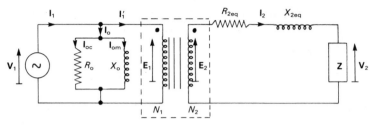

Figure 5.8 Approximate equivalent circuit with series components referred to secondary side

$$R_{2eq} = R_1' + R_2 = R_1 \left\{ \frac{N_2}{N_1} \right\}^2 + R_2 \qquad (5.19)$$

and

$$X_{2eq} + X_1' + X_2 = X_1 \left\{ \frac{N_2}{N_1} \right\}^2 + X_2 \qquad (5.20)$$

When referring impedances, the rule is that the side to which the impedance is referred has its winding turns 'on top' in the turns-ratio-squared expression.

The technique may be extended to include the load, if its value is known, and the transformer eliminated altogether as indicated in Figure 5.9. The transformer model is then reduced to a simple electrical network with no induced voltages present.

Transformers may be used for impedance matching; the matching of load impedance to source impedance. The standard application of this principle has been the matching of a loudspeaker load to the output impedance of an amplifier in audio circuits. It was shown in Chapter 3 how maximum power is transferred from a source to a resistive load when source resistance is equal to load resistance. It is evident that, for a resistive load R_L connected to a source with resist-

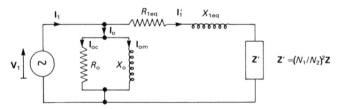

Figure 5.9 Approximate equivalent circuit with all series components, including load, referred to primary side

ive source impedance R_s, matching may be achieved by inserting a transformer with turns ratio such that:

$$R_s = \left\{\frac{N_1}{N_2}\right\}^2 R_L \tag{5.21}$$

5.5 Transformer tests

The equivalent circuits developed facilitate performance prediction for transformers under normal operating conditions. However, to use them, values are required for the various circuit components. Two straightforward tests may be performed to determine these values for any transformer.

The *open-circuit test* consists of measurement of voltage, current, and power on the primary side and voltage on the secondary side, with the secondary open-circuited. With reference to the approximate equivalent circuit of Figure 5.8, it is evident that both I_1' and I_2 are zero in this case. The equivalent circuit therefore reduces to the form shown in Figure 5.10(a), in which only the parallel components R_o and X_o remain. The turns ratio follows initially from the ratio of the voltage readings. If the measured values of voltage, current, and

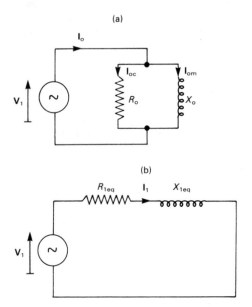

Figure 5.10 Equivalent circuits for transformer tests   (a) open circuit   (b) short circuit

power on the primary side are V_1, I_o, and P respectively, then from Figures 5.10(a) and 5.3(b)

$$P = V_1 I_o \cos \varphi_o \tag{5.22}$$

$$R_o = \frac{V_1}{I_{oc}} = \frac{V_1}{I_o \cos \varphi_o} = \frac{V_1^2}{P} \tag{5.23}$$

and

$$X_o = \frac{V_1}{I_{om}} = \frac{V_1}{I_o \sin \varphi_o} = \frac{V_1^2}{\sqrt{(V_1^2 I_o^2 - P^2)}} \tag{5.24}$$

If the transformer is subsequently used with primary and secondary connections interchanged, then R_o and X_o may be referred in the same manner as the series components by the turns-ratio-squared factor.

The complementary *short-circuit test* is performed by short-circuiting the secondary winding, and supplying the primary with just sufficient voltage to circulate a current of the order of the full-load current through the transformer. Measurement of primary voltage, current and power is taken as before. The precise current level at which the test is carried out is unimportant; full-load current is normally used because this is usually the load condition of greatest interest. The vital aspect of the test is the short-circuit, the significance of which should be appreciated by reference again to Figure 5.7, which shows the approximate equivalent circuit with the series components referred to the primary side. On short-circuit, $V_2 = 0$ which implies both E_2 and E_1 must be zero in this circuit, since there are no circuit components on the secondary side. Hence the circuit reduces to the no-load components R_o, X_o in parallel with the series components referred to the primary side. Due to the much lower relative impedance of these series components, negligible current flows through the R_o/X_o parallel combination in this case. R_o and X_o may consequently be neglected, and the equivalent circuit reduces to the series components alone, as shown in Figure 5.10(b). If the primary values of voltage, current, and power measured on the short-circuit test are V_1, I_1, and P respectively, then:

$$R_{1eq} = P/I_1^2 \tag{5.25}$$

and

$$X_{1eq} = \frac{\sqrt{(V_1^2 I_1^2 - P^2)}}{I_1^2} \tag{5.26}$$

The short-circuit test may be carried out on either side of the transformer, depending on which is most suited to the supply available.

The series component values obtained are referred to the supply (primary) side. It is not possible to separate the primary and secondary series components from the results of the test. However, it is not necessary to do this to model the transformer behaviour, since, as has been demonstrated, this can be readily achieved with all series components referred to the same side.

5.6 Transformer performance

When considering transformer operation, it is imperative to have a means of specifying the maximum load or output that it can supply. A transformer is usually designed to operate between predetermined voltage levels and at a known fixed frequency. Numbers of turns and core cross-sectional area are then related by the peak flux density allowable from iron loss and magnetising current considerations. The maximum output of the transformer is then essentially limited by its heat dissipation capabilities. The iron loss is assumed constant, and the winding power loss increases with the load and supply currents. Hence a maximum current may be specified which the transformer can supply continuously without overheating. This information is expressed by the *VA* (or *KVA* or *MVA*) *rating* of the transformer. The full load primary or secondary currents may then be obtained by dividing the rating by the corresponding rated voltage. As an example, a 500 MVA, 400/132 kV transformer has full load currents of 1250 A and 3790 A on its high and low voltage sides respectively.

It is important to note that the rating is not, in general, equal to the full load power transformed. The rating is independent of the phase of the current. The operating power factor is fixed by the power factor of the load connected to the secondary. Only when supplying full-load current to a unity-power-factor load will the transformer's power be equal to its rating.

Another important transformer characteristic is its *voltage regulation*. This is a measure of the change in the secondary terminal voltage as the load, and current, varies. In small transformers used in electronic circuits, this voltage variation may not be critical. However, in power transformers used in supply and distribution systems, voltage levels must be maintained fairly constant, and tap-changing techniques are used to vary winding turns in order to achieve this.

From the equivalent circuits, it is evident that regulation is caused by voltage drops in winding resistance and winding leakage reactance. Its magnitude may be readily calculated from the approximate equivalent circuit of Figure 5.8. This version of the circuit is most convenient, since the secondary induced voltage $\mathbf{E}_2$ remains constant. The corresponding phasor diagram for the secondary circuit is shown in Figure 5.11. An inductive load with lagging phase angle φ is assumed. The circuit equation is

$$\mathbf{E}_2 = \mathbf{V}_2 + (R_{2eq} + jX_{2eq})\mathbf{I}_2 \tag{5.27}$$

To calculate regulation, this equation must be rearranged to yield the difference in magnitude of $\mathbf{E}_2$ and $\mathbf{V}_2$. This is most easily achieved by equating components in phase and quadrature with $\mathbf{V}_2$ i.e.

$$E_2 \cos \delta = V_2 + (R_{2eq} \cos \varphi + X_{2eq} \sin \varphi)I_2 \tag{5.28}$$

$$E_2 \sin \delta = (X_{2eq} \cos \varphi - R_{2eq} \sin \varphi)I_2 \tag{5.29}$$

where δ is the phase angle between $\mathbf{E}_2$ and $\mathbf{V}_2$. The general solution may then be obtained by summing the squares of both sides to eliminate δ. In practice, regulation is small and little accuracy is lost by assuming $\delta = 0$. Equation (5.29) then becomes zero and the regulation follows from Equation (5.28) as:

$$E_2 - V_2 = (R_{2eq} \cos \varphi + X_{2eq} \sin \varphi)I_2 \tag{5.30}$$

With the lagging-power-factor load assumed, the regulation is positive and the secondary voltage falls as the load current increases. In the unusual event of a capacitive load and leading power factor, φ and $\sin \varphi$ become negative in Equation (5.30) and the secondary voltage may increase with load current, giving negative regulation.

It is more informative to express the regulation as a proportion or percentage of the no-load secondary voltage or the full-load secondary voltage. In practice, regulation is only of the order of a few percent and the value is little affected by which of these voltages is used as a reference in the calculation.

Transformer efficiency may be readily calculated from the equivalent circuit. In the approximate equivalent circuit, the iron loss is evidently constant and of value $R_o I_{oc}^2$ or V_1^2/R_o. For the approximate equivalent circuit with components referred to the secondary (see

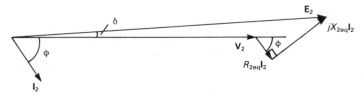

Figure 5.11 Secondary phasor diagram with series components referred to primary side

Figure 5.8), the winding copper loss is $R_{2eq}I_2^2$. Then for a load with power factor $\cos \varphi$:

$$\text{Efficiency } (\eta) = \frac{\text{output power}}{\text{output power} + \text{losses}}$$

$$= \frac{V_2 I_2 \cos \varphi}{V_2 I_2 \cos \varphi + R_{2eq}I_2^2 + V_1^2/R_o} \qquad (5.31)$$

On no-load, I_2 and efficiency are zero. As load and I_2 increase, efficiency increases and a condition of maximum efficiency is achieved. This condition for a constant-power-factor load may be determined by differentiation of Equation (5.31) with respect to I_2 making the reasonable assumption in these circumstances that V_2 is constant, or that regulation may be neglected. Setting the differential equal to zero gives the condition for maximum efficiency to be:

$$V_1^2/R_o = R_{2eq}I_2^2 \qquad (5.32)$$

or that *the winding copper loss must equal the iron loss*.

Transformers are very efficient devices, with maximum efficiencies varying typically from 90% in the smallest sizes, up to and exceeding 99% in the largest sizes manufactured (currently ratings of 700 MVA or more). It is instructive to consider why transformer efficiency increases with size. This may be justified by the following somewhat superficial argument which nevertheless gives substantially the correct result. Figure 5.12 shows two transformers having the same turns ratio and connected to the same supply. They are assumed to be connected to loads of identical power factor. The second transformer is bigger than the first in the ratio $x:1$ in all its linear dimensions. This implies that areas are increased in the ratio $x^2:1$ and volumes in the ratio $x^3:1$. If both are designed to make best use of the core material, then they must operate with identical peak magnetic flux densities. Hence, peak flux is increased in the ratio x^2, and so winding turns must be reduced in the ratio $1/x^2$ to satisfy the transformer Equation (5.5), since the supply voltage is unchanged. The effect of this change on winding resistance requires careful consideration, and it is best considered in two stages. The argument is based on the basic resistance relationship for a conductor, given as Equation (2.18). The resistance of an N-turn coil of constant cross-sectional-area varies as N^2, since coil length is proportional to N, and the cross-sectional-area of each individual turn is inversely proportional to N. The effect of the $x:1$ size change on a coil with a fixed number of turns is to reduce its resistance in the ratio $1/x$. Hence the winding resistance of the bigger transformer with its reduced number of turns will be reduced in the ratio $1/x^5$.

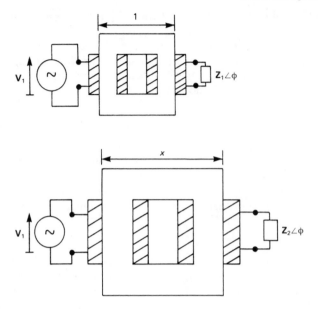

Figure 5.12 Effect of transformer size on performance

The losses of the transformers may now be considered. From Equations (4.11) for hysteresis loss and (4.15) for eddy-current loss, it should be apparent that the specific iron loss is independent of the size. Hence the actual iron loss of the bigger transformer will be increased in the ratio x^3. In order to maintain iron loss and copper loss of comparable magnitude, it is reasonable to assume that I^2R in the transformer may be increased in the same ratio. With R reduced in the ratio $1/x^5$, it follows that I must be increased in the ratio x^4:1 to satisfy this criterion.

The effect on rating, power output, and efficiency should now become apparent. Voltage levels are unchanged, and so with a load of constant power factor, both rating and power output are increased in the current ratio of x^4:1. However, losses have only increased in the ratio x^3:1. Hence, the ratio of losses to power output and rating has decreased, giving increased efficiency.

The influence of size on transformer cooling may also be readily explained. The losses must be removed in the form of heat, which is facilitated by having the maximum possible surface area available. With losses increasing as x^3 but surface area increasing only as x^2, cooling clearly represents a problem which grows with size. This is why small transformers can rely on natural air circulation for sufficient cooling, whereas large power transformers require separate free-standing banks of radiators, with closed oil circulation systems forced by pumps, and additional fan cooling.

This important general principle, that efficiency and difficulty in cooling increase with size applies not only to transformers but also to electromechanical-energy-conversion equipment.

5.7 The autotransformer

The autotransformer is a particular type of transformer in which the secondary winding is dispensed with and the secondary supply is taken from between one common end of the primary winding, and a tapping some point along its length. The arrangement is shown in Figure 5.13(a). If ideal transformer properties are assumed, and the

(a)

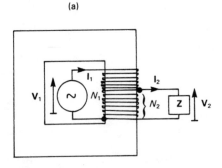

(b)

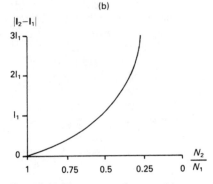

Figure 5.13 The autotransformer   (a) connection   (b) variation in current in common part of winding

secondary taps off N_2 turns of the total N_1 turns in the winding, then

$$\frac{V_1}{V_2} = \frac{N_1}{N_2} \tag{5.33}$$

as in the conventional transformer.

When load is connected, I_1 and I_2 may be related by the principle of Ampère-turn balance; the nett m.m.f. set up in the core by the load currents must be zero. This gives:

$$I_1(N_1 - N_2) + (I_1 - I_2)N_2 = 0$$

or

$$N_1 I_1 = N_2 I_2 \tag{5.34}$$

which again is the identical result to that obtained with a conventional transformer.

The current in the lower (common) part of the winding is:

$$I_2 - I_1 = I_1(N_1/N_2 - 1) \tag{5.35}$$

The variation of this current with turns ratio N_2/N_1 is shown in Figure 5.13(b). For a 2:1 step down in voltage the current in the common part is equal in magnitude to the current in the top part. Hence, the device operates as a conventional 2:1 step-down transformer but no secondary winding is required, making the transformer smaller and less expensive. However, it is evident that for voltage ratios greater than 2:1 the current in the common part rapidly increases over that in the top part. This would require increased conductor cross-sectional area and the benefit of the connection would quickly be lost. In addition, the autotransformer has the added disadvantage that electrical isolation between primary and secondary is also lost. Hence its main application is in systems where the voltage ratio is not too great and where electrical isolation is not essential. It is commonly used to perform the 275/132 kV voltage change in the UK power transmission network. In small sizes, it is also frequently encountered in electrical laboratories in a cylindrical form with movable brushes to provide variable low-voltage a.c. supplies.

5.8 Practical transformer construction

In Figure 5.1(a) and subsequent similar diagrams, the transformer windings are shown wound on separate limbs to clarify the flux paths associated with each. This arrangement results in excessive flux

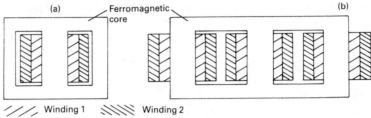

Figure 5.14 Practical transformer arrangements   (a) one phase   (b) three phase

leakage and so, in practice, the windings are normally placed, one on top of the other, as indicated in Figure 5.14(a).

In power-transmission systems, three sinusoidal a.c. supplies are commonly used, each of equal magnitude but mutually equispaced in time phase. Correspondingly, *three-phase* transformers are used to change voltage levels in such systems, requiring three pairs of windings, often arranged as shown in Figure 5.14(b). For further details of three-phase systems, the interested reader is referred to the appropriate companion volume (Laughton and Adams).

5.9 Bibliography

LAUGHTON, M. A. and ADAMS, R. N., *BASIC Electrical Power Systems*, Butterworths, London (in press)

WORKED EXAMPLES

Example 5.1: TRNSTNS: calculation of transformer turns

Develop a program to calculate the numbers of turns required on a two-winding transformer. Input data is to consist of voltage levels, supply frequency, and cross-sectional area and peak magnetic flux density in the transformer core.

```
10 REM      TRNSTNS - Calculation of transformer turns
20 P1 = 3.14159265
30 R2 = SQR(2)
40 PRINT
50 PRINT
60 PRINT "Calculation of transformer turns"
70 PRINT "--------------------------------"
80 PRINT
90 PRINT "Voltage on L.V. side ";
100 INPUT V1
110 PRINT
120 PRINT "Voltage on H.V. side ";
130 INPUT V2
140 PRINT
```

```
150 PRINT "Supply frequency (Hz) ";
160 INPUT F
170 PRINT
180 PRINT "Core cross-section-area (cm^2) ";
190 INPUT A
200 PRINT
210 PRINT "Peak flux density allowed (T) ";
220 INPUT B
230 P = B*A*1E-4
240 N1 = V1/(P1*R2*F*P)
250 N1 = INT(N1 + 1)
260 N2 = N1*V2/V1
270 N2 = INT(N2 + 0.5)
280 PRINT
290 PRINT
300 PRINT " Low voltage winding has "; N1; " turns"
310 PRINT
320 PRINT "High voltage winding has "; N2; " turns"
330 PRINT
340 PRINT
350 PRINT "Another calculation (Y/N) ";
360 INPUT Q$
370 IF Q$ = "Y" THEN GOTO 40
380 STOP
390 END
```

```
Calculation of transformer turns
---------------------------------

Voltage on L.V. side ?10

Voltage on H.V. side ?240

Supply frequency (Hz) ?50

Core cross-section-area (cm^2) ?1

Peak flux density allowed (T) ?1.6

 Low voltage winding has 282 turns

High voltage winding has 6768 turns

Another calculation (Y/N) ?N

STOP at line 380
```

Program notes

(1) Equation (5.5) forms the basis for the program. Peak flux density (P) is evaluated at line 230, and the exact value of turns to give the lower voltage at the peak allowable flux density is assigned to N1 at the following line. Since the turns number must be an integer, the INT

function is used to convert N1 to the smallest integer in excess of this value (line 250). This procedure ensures that the transformer operates at the highest possible flux density not exceeding the allowed maximum input.

(2) The turns for the higher voltage then follow from the ratio of voltage levels (line 260). Again, this must be an integer, and so N2 is rounded to the nearest whole number using the INT function (line 270).

Example 5.2: IOMAG: magnetising current waveform

It has been explained how magnetically non-linear iron results in a non-sinusoidal magnetising current, I_{om}, setting up the sinusoidal flux. Develop a program to demonstrate the actual shape of I_{om} waveform required. Core material is to be cold-rolled, grain-oriented steel with a **B–H** characteristic as shown in Figure 4.2. Input data required is supply frequency, transformer primary winding turns and voltage, and length and cross-sectional area of the iron circuit. An allowance should also be made for any unwanted airgaps introduced into the iron circuit by the manufacturing process.

```
10 REM      IOMAG - Magnetising current waveform
20 DATA 0,0, 0.65,50, 1.3,100, 1.52,200, 1.63,300
30 DATA 1.7,500, 1.76,1000, 1.84,2000, 1.89,3000
35 DATA 1.94,5000, 2,8000
40 DIM I1(10,1)
50 FOR J = 0 TO 10
60    FOR K = 0 TO 1
70       READ I1(J,K)
80       NEXT K
90    NEXT J
100 P1 = 3.14159265
110 MO = 4*P1*1E-7
120 R2 = SQR(2)
130 PRINT
140 PRINT
150 PRINT "Magnetising current (zero hysteresis)"
160 PRINT "-------------------------------------"
170 PRINT "(cold-rolled grain-oriented steel)"
180 PRINT
190 PRINT "Transformer primary turns ";
200 INPUT N
210 PRINT
220 PRINT "Rated r.m.s. primary voltage (V) ";
230 INPUT E
240 PRINT
250 PRINT "Supply frequency (Hz) ";
260 INPUT F
270 PRINT
280 PRINT "Effective length of iron circuit (m) ";
290 INPUT L1
300 PRINT
```

```
310 PRINT "Overall airgap in iron circuit (mm) ";
320 INPUT L2
330 PRINT
340 PRINT "Core cross-sectional-area (cm^2) ";
350 INPUT A
360 P2 = E/(R2*P1*F*N)
370 B1 = P2/(A*1E-4)
380 IF B1>2.0 THEN GOTO 630
390 K1 = L1/N
400 K2 = L2*1E-3/(M0*N)
410 PRINT
420 PRINT "Peak core flux density "; B1; " T"
430 PRINT
440 PRINT "Angle (deg)      Current (A)"
450 PRINT "------------     ------------"
460 PRINT
470 FOR T = 0 TO 20
480    B = B1*SIN(T*P1/20)
490    FOR J = 0 TO 10
500      IF I1(J,0) > B THEN GOTO 520
510      NEXT J
520    X1 = I1(J,1) - I1(J-1,1)
530    Y1 = I1(J,0) - I1(J-1,0)
540    H = I1(J-1,1) + (B - I1(J-1,0))*X1/Y1
550    I0 = K1*H + K2*B
560    PRINT TAB(4); 9*T; TAB(17); I0
570    NEXT T
580 PRINT
590 PRINT "Another run (Y/N) ";
600 INPUT Q$
610 IF Q$ = "Y" THEN GOTO 130
620 STOP
630 PRINT
640 PRINT "Peak flux density too high -"
650 PRINT "increase turns or core area "
660 GOTO 130
670 END


>RUN


Magnetising current (zero hysteresis)
-------------------------------------
(cold-rolled grain-oriented steel)

Transformer primary turns ?1420

Rated r.m.s. primary voltage (V) ?240

Supply frequency (Hz) ?50

Effective length of iron circuit (m) ?.2

Overall airgap in iron circuit (mm) ?1

Core cross-sectional-area (cm^2) ?4

Peak core flux density 1.902 T
```

Angle (deg)	Current (A)
0	0
9	0.17
18	0.3358
27	0.4933
36	0.6387
45	0.7707
54	0.8929
63	1.018
72	1.241
81	1.443
90	1.557
99	1.443
108	1.241
117	1.018
126	0.8929
135	0.7707
144	0.6387
153	0.4933
162	0.3358
171	0.17
180	4.18E-9

Another run (Y/N) ?Y

Magnetising current (zero hysteresis)

(cold-rolled grain-oriented steel)

Transformer primary turns ?1420

Rated r.m.s. primary voltage (V) ?240

Supply frequency (Hz) ?50

Effective length of iron circuit (m) ?.2

Overall airgap in iron circuit (mm) ?.1

Core cross-sectional-area (cm^2) ?4

Peak core flux density 1.902 T

Angle (deg)	Current (A)
0	0
9	1.99E-2
18	3.931E-2
27	5.775E-2
36	7.477E-2
45	9.234E-2
54	0.1168
63	0.1633
72	0.3285
81	0.4959
90	0.5972
99	0.4959
108	0.3285
117	0.1633
126	0.1168
135	9.234E-2
144	7.477E-2
153	5.775E-2
162	3.931E-2
171	1.99E-2
180	4.893E-10

```
Another run (Y/N) ?N

STOP at line 620
```

Program notes

(1) The **B–H** data for the iron is held in a similar manner to that employed in Examples 4.1 and 4.2.

(2) Peak flux (P2) is evaluated at line 360 using Equation (5.5). Peak flux density (B1) follows directly at the subsequent line. A check is made at line 380 that peak flux density does not exceed the maximum value stored.

(3) I_{om} is calculated at 9° intervals using Equation (5.8). Two constants K1 and K2 are first calculated at lines 390 and 400, related to the iron and airgap components of the current. At each interval, the value of **B** is calculated from knowledge of its sinusoidal variation and peak value (line 480). Linear interpolation is then used to calculate the corresponding **H** for the iron, in an identical manner to that employed in Example 4.2 (lines 490–540). Instantaneous current then follows directly at line 550.

(4) The two runs illustrate the effect of manufacturing quality on transformer magnetising current. The data is identical, excepting for the airgap in the iron circuit. The first (1 mm airgap) shows a high current value, but limited harmonic content due to the considerable m.m.f. dropped across the magnetically-linear airgap. The second (0.1 mm airgap) implying higher quality manufacture, has much reduced overall current level. However, the distortion is now much

increased since the magnetically non-linear iron dominates the characteristics obtained. The results for one cycle are shown graphically in Figure 5.15.

Example 5.3: TRANSCT: 'exact' equivalent circuit for transformer

Write a program to solve the 'exact' equivalent circuit of the two-winding transformer (Figure 5.5). Fixed frequency operation may be assumed, and so ohmic reactance values may be input. Results displayed should include winding currents and induced e.m.fs, input power factor, component and total power losses, input and output powers, and efficiency.

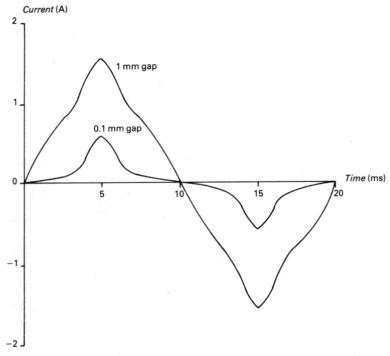

Figure 5.15 Magnetising current waveforms

```
10 REM        TRANSCT - Transformer performance using
20 REM                   'exact' equivalent circuit
30 PRINT
40 PRINT
50 PRINT "Transformer performance - 'exact' equivalent";
```

```
 55 PRINT " circuit"
 60 PRINT "-----------------------------------------------";
 65 PRINT "---------"
 70 PRINT
 80 PRINT  "Resistance - primary winding (Ohms) ";
 90 INPUT R1
100 PRINT
110 PRINT "Leakage reactance - primary winding (Ohms) ";
120 INPUT X1
130 PRINT
140 PRINT "Iron loss resistance (Ohms) ";
150 INPUT R0
160 PRINT
170 PRINT "Magnetising reactance (Ohms) ";
180 INPUT X0
190 PRINT
200 PRINT "Resistance - secondary winding (Ohms) ";
210 INPUT R2
220 PRINT
230 PRINT "Leakage reactance - secondary winding (Ohms) ";
240 INPUT X2
250 PRINT
260 PRINT "Turns ratio (N2/N1) ";
270 INPUT N
280 PRINT
290 PRINT
300 PRINT "Supply (primary) voltage ";
310 INPUT V1
320 PRINT
330 PRINT "Load resistance (Ohms) ";
340 INPUT R3
350 PRINT
360 PRINT "Load reactance (N.B. cap. reactance is -ve)";
365 PRINT " (Ohms) ";
370 INPUT X3
380 A = N^2
390 B = 0
400 C = R2 + R3
410 D = X2 + X3
420 GOSUB 1280
430 A = 1
440 B = 0
450 C = 1/R0 + D1
460 D = -1/X0 + D2
470 GOSUB 1280
480 A = V1
490 B = 0
500 C = R1 + D1
510 D = X1 + D2
520 GOSUB 1280
530 I1 = D1
540 I6 = D2
550 A = R1
560 B = X1
570 C = I1
580 D = I6
590 GOSUB 1200
600 E1 = V1 - M1
610 E6 = -M2
```

```
620 A = E1
630 B = E6
640 C = 1/R0
650 D = -1/X0
660 GOSUB 1200
670 I0 = M1
680 I5 = M2
690 I3 = I1 - I0
700 I8 = I6 - I5
710 I2 = I3/N
720 I7 = I8/N
730 A = R3
740 B = X3
750 C = I2
760 D = I7
770 GOSUB 1200
780 V2 = M1
790 V7 = M2
800 P = ATN(I6/I1)
810 P3 = COS(P)
820 PRINT
830 PRINT
840 PRINT "Input power factor "; P3;
850 IF I6>0 THEN PRINT " leading"
860 IF I6<0 THEN PRINT " lagging"
870 PRINT
880 PRINT "  Primary current "; SQR(I1^2 + I6^2); " A"
890 PRINT "Secondary current "; SQR(I2^2 + I7^2); " A"
900 PRINT
910 PRINT "        Primary induced e.m.f. ";
915 PRINT SQR(E1^2 + E6^2); " V"
920 PRINT "      Secondary induced e.m.f. ";
925 PRINT N*SQR(E1^2 + E6^2); " V"
930 PRINT "Terminal (Secondary) voltage ";
935 PRINT SQR(V2^2 + V7^2); " V"
940 PRINT
950 P0 = (E1^2 + E6^2)/R0
960 P1 = (I1^2 + I6^2)*R1
970 P2 = (I2^2 + I7^2)*R2
980 PRINT "          Iron loss "; P0; " W"
990 PRINT " Primary winding loss "; P1; " W"
1000 PRINT "Secondary winding loss "; P2; " W"
1010 PRINT "                       ------------"
1020 PRINT "            Total loss "; P0 + P1 + P2; " W"
1030 PRINT "                       ------------"
1040 PRINT
1050 P3 = (I2^2 + I7^2)*R3
1060 PRINT " Input power "; P3 + P0 + P1 + P2; " W"
1070 PRINT "Output power "; P3; " W"
1080 PRINT
1090 PRINT "Efficiency "; 100*P3/(P3 + P0 + P1 + P2); " %"
1100 PRINT
1110 PRINT "Another run with different voltage or load ";
1120 INPUT Q$
1130 IF Q$ = "Y" THEN GOTO 280
1140 STOP
1150 REM
1160 REM *** COMPLEX multiplication subroutine ***
1170 REM
1180 REM M1 + jM2 = (A + jB)*(C + jD)
1190 REM
1200 M1 = A*C - B*D
```

```
1210 M2 = A*D + B*C
1220 RETURN
1230 REM
1240 REM *** COMPLEX division subroutine ***
1250 REM
1260 REM D1 + jD2 = (A + jB)/(C + jD)
1270 REM
1280 D0 = C^2 + D^2
1290 D1 = (A*C + B*D)/D0
1300 D2 = (B*C - A*D)/D0
1310 RETURN
1320 END
```

Transformer performance - 'exact' equivalent circuit

Resistance - primary winding (Ohms) ?.0715

Leakage reactance - primary winding (Ohms) ?.1825

Iron loss resistance (Ohms) ?1300

Magnetising reactance (Ohms) ?325

Resistance - secondary winding (Ohms) ?.00387

Leakage reactance - secondary winding (Ohms) ?.00872

Turns ratio (N2/N1) ?.2182

Supply (primary) voltage ?1100

Load resistance (Ohms) ?.5

Load reactance (N.B. cap. reactance is -ve) (Ohms) ?.25

Input power factor 0.87141 lagging

   Primary current 93.579 A
Secondary current 418.31 A

      Primary induced e.m.f. 1085.9 V
    Secondary induced e.m.f. 236.93 V
Terminal (Secondary) voltage 233.84 V

             Iron loss 906.98 W
  Primary winding loss 626.13 W
Secondary winding loss 677.17 W

            Total loss 2210.3 W

 Input power 89700 W
Output power 87490 W

Efficiency 97.536 %

Another run with different voltage or load ?N

STOP at line 1140

Program notes

(1) Although the length of the program may be somewhat intimidating, its content should not be difficult to understand, being merely based on simple circuit theory and referred impedances. The complex multiplication and division required is performed by subroutines beginning at lines 1150 and 1230 respectively. The explanation should be followed with the aid of Figure 5.5.

(2) The first principal quantity evaluated is I_1, obtained by reducing the whole circuit to a single overall impedance. The transformer is 'eliminated' by referring the total secondary admittance Y_2 (winding in series with load) to the primary side (lines 380–420). This referred admittance Y_2' then carries current I_1' and appears in parallel with the no-load branch of the circuit, replacing the transformer and the secondary circuit. The total impedance of the no-load branch and Y_2' then follows from the reciprocal of their admittances added in parallel (lines 430 to 470). The overall impedance is obtained from the series sum of the primary winding impedance and this impedance. This overall impedance, divided into V_1, gives I_1 (lines 480 to 540). Real and imaginary parts of I_1 are assigned to I1 and I6 respectively. Mathematically, the above process is represented by the equation

$$I_1 = \frac{V_1}{R_1 + jX_1 + 1/(Y_o + Y_2')} \tag{5.36}$$

where

$$Y_o = 1/R_o - j/X_o$$

and

$$Y_2' = \left\{\frac{N_1}{N_2}\right\}^2 \cdot \frac{1}{R_2 + jX_2 + Z}$$

(3) The voltage drop $I_1 (R_1 + jX_1)$ is evaluated next (lines 550–590). Subtraction of this from V_1 gives E_1, with real and imaginary parts assigned to E1 and E6 respectively (lines 600 and 610).

(4) I_o is obtained from $Y_o E_1$, with real and imaginary parts I0 and I5 (lines 620–680). Subtraction of I_o from I_1 leaves I_1' (lines 690 and 700). I_2 then follows from the turns ratio (lines 710 and 720).

(5) ZI_2 gives V_2 directly (lines 730–790) with real and imaginary parts assigned to V2 and V7 respectively.

(6) The methods used to calculate the remaining quantities from the voltages and currents should be self-evident. In particular, the power factor follows from the phase angle of I_1. Losses are given by the power dissipation in corresponding equivalent circuit resistances, and the output power is the power associated with the load resistance.

(7) The sample run is for a 100 kVA, 50 Hz, 1100/240 V transformer

operating under approximately full-load conditions. Note the small difference between V_1 and E_1, showing that little error would be introduced if the 'approximate' equivalent circuit were used in this case. The efficiency is high as expected, and it is possible, by varying the load, to verify that maximum efficiency occurs with a resistive load when the iron loss is half the total loss.

Example 5.4: TRANTST: equivalent circuit parameters from test results

Develop a program to process the results of open-circuit and short-circuit transformer tests.

```
10 REM      TRANTST - Equiv. circuit parameters from
20 REM      results of open and short-circuit tests
30 PRINT
40 PRINT
50 PRINT "Equivalent circuit parameters from test";
55 PRINT " results"
60 PRINT "--------------------------------------------";
65 PRINT "--------"
70 PRINT
80 PRINT "Open-circuit test"
90 PRINT "------------------"
100 PRINT
110 PRINT "Primary (supply) voltage ";
120 INPUT V1
130 PRINT
140 PRINT "O.C. Secondary voltage ";
150 INPUT V2
160 PRINT
170 PRINT "No-load Primary current ";
180 INPUT I0
190 PRINT
200 PRINT "No-load supply power (W) ";
210 INPUT P0
220 PRINT
230 PRINT
240 PRINT "Turns ratio (N2/N1) = "; V2/V1
250 S = V1^2
260 PRINT
270 PRINT "R0 = "; S/P0; " Ohms (on supply side)"
280 PRINT
290 PRINT "X0 = "; S/SQR(S*I0^2 - P0^2);
295 PRINT " Ohms (on supply side)"
300 PRINT
310 PRINT
320 PRINT "Short-circuit test"
330 PRINT "------------------"
340 PRINT
350 PRINT "Primary (supply) voltage ";
360 INPUT V1
370 PRINT
380 PRINT "Primary current ";
390 INPUT I1
```

```
400 PRINT
410 PRINT "Short-circuit supply power (W) ";
420 INPUT P1
430 S = I1^2
440 PRINT
450 PRINT
460 PRINT "R1eq = "; P1/S; " Ohms (on supply side)"
470 PRINT
480 PRINT "X1eq = "; SQR(V1^2*S - P1^2)/S;
485 PRINT " Ohms (on supply side)"
490 PRINT
500 PRINT
510 PRINT "Results for another transformer (Y/N) ";
520 INPUT Q$
530 IF Q$ = "Y" THEN GOTO 30
540 STOP
550 END
```

```
Equivalent circuit parameters from test results
-----------------------------------------------

Open-circuit test
-----------------

Primary (supply) voltage ?240

O.C. Secondary voltage ?1100

No-load Primary current ?16

No-load supply power (W) ?931


Turns ratio (N2/N1) = 4.583

RO = 61.87 Ohms (on supply side)

XO = 15.46 Ohms (on supply side)


Short-circuit test
------------------

Primary (supply) voltage ?7.9

Primary current ?420

Short-circuit supply power (W) ?1280


R1eq = 7.256E-3 Ohms (on supply side)

X1eq = 1.735E-2 Ohms (on supply side)


Results for another transformer (Y/N) ?N

STOP at line 540
```

Program notes
(1) The program follows the theory of Section 5.5. Equations (5.23) and (5.24) are implemented at lines 270 and 290 respectively. Likewise, Equations (5.25) and (5.26) follow at lines 460 and 480 respectively.
(2) The results shown are for the transformer previously considered in Example 5.3. Checking the consistency of the equivalent-circuit parameters is left as an exercise for the student.

PROBLEMS

(5.1) Test results on a quantity of cold-rolled, grain-oriented steel laminations showed them to have a specific iron loss at 50 Hz of 2.0 W/kg with a sinusoidally-varying flux density of peak value 1.5 T. For convenience, assume the hysteresis and eddy-current loss components to be equal at this frequency, and the Steinmetz coefficient to be 2. Taking the specific gravity of the laminations to be 7.7, extend program IOMAG of Example 5.2 to calculate the approximate shape of the total no-load current I_o. Include also an estimation of the no-load power factor.

What extra information would be required to improve the accuracy of the predicted waveshape? What complications are involved in the calculation of power factor? Use the original IOMAG data to investigate the effect of manufacturing quality on no-load power factor.

(Hint: Begin by examining Equation (4.24) and Equation (5.9) with its adjacent text.)

(5.2) Develop a program to solve the 'approximate' equivalent circuit of a transformer, shown in Figure 5.6. The same basic approach and layout may be employed as for Example 5.3 (TRANSCT). Identical data may then be run for each program to check the differences between the 'exact' and 'approximate' models.

(5.3) In power applications, the load on a transformer is more often known or specified in terms of the VA (or kVA or MVA) available at its secondary terminals. Produce modified versions of TRANSCT and the previous program to cater for this. Revised input data should now include secondary terminal (load) VA, load power factor, and secondary terminal voltage, in place of supply voltage, and load resistance and reactance.

If secondary terminal voltage is held constant, and the modified programs run with increasing load VA at a fixed lagging power factor, the predicted supply voltage should be found to increase. Transformers in power systems are normally required to operate between fixed voltage levels. How is this difficulty overcome in such a transformer in practice?

(5.4) It is instructive to establish how voltage regulation varies with load current and power factor. Write a program to do this, using Equations (5.28) and (5.29) as indicated in the text. Input data is to be primary and secondary winding resistances and leakage reactances, and transformer rated secondary terminal voltage.

(6.5) The theoretical effects of size on transformer performance are explored at the end of Section 5.6. Use the results to develop a program as follows: the rating of a transformer and its maximum efficiency are input, together with the rating of a second transformer. Output the predicted linear scale relationship of the second transformer to the first, together with its expected maximum efficiency.

(Note: The results of a program of this type should only be considered as a rough guide to the general trend. Scaling techniques should always be regarded with caution, especially when carried out over a wide range, since constructional details may well differ, rendering some of the basic assumptions invalid.)

Electromechanical energy conversion

ESSENTIAL THEORY

6.1 Introduction

A study of the interchange of energy in electromechanical systems is fundamental to the understanding of how electrical machines work. Application of the Principle of Conservation of Energy allows forces to be calculated for straight-line motion and the corresponding torques to be evaluated in rotary systems. Force-producing devices based on electromagnetics are far more commonly encountered than corresponding ones using electrostatic effects, and the reason for this becomes apparent when relative force levels are compared.

This chapter considers magnetically-linear systems in some detail, where results may generally be obtained in the form of simple formulae. The discussion is then widened to include non-linear systems, where the idea of field co-energy is introduced to aid the analysis. Non-linearity does not significantly increase the conceptual difficulty, but more laborious graphical techniques are often required to obtain solutions.

6.2 Energy balance in electromechanical system

Four distinct forms of energy may be identified as being present in electromechanical systems:

(1) Electrical energy (W_e) which is associated with the electrical supply.
(2) Mechanical energy (W_m) which corresponds to work done by force × distance moved or torque × angle turned through.
(3) Energy (W_f) stored in the magnetic (or electric) field.
(4) Energy losses, which include electrical (copper loss and eddy-current loss), mechanical (friction and windage loss) and magnetic (hysteresis) losses.

Any interchange of electrical and mechanical energy takes place via the stored energy of the field. In an analysis of this process, the losses may be ignored since these occur externally to the field. A small inter-

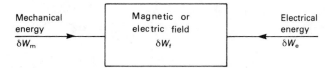

Figure 6.1 The electromechanical energy conversion process

change of energy may hence be represented by the system shown in Figure 6.1. The convention adopted is that energy supplied to the field is considered positive. Machines may be classified by the direction of energy transfer.

An *electric motor* converts electrical energy into mechanical energy. Electrical energy is supplied to the field, and mechanical energy is extracted from the field. With our convention, δW_e is positive and δW_m is negative.

An *electric generator* converts mechanical energy into electrical energy. Mechanical energy is supplied to the field, and electrical energy is extracted from it. δW_e is negative and δW_m is positive.

In either case, the following energy balance applies for the energy interchange:

$$\delta W_e + \delta W_m = \delta W_f \tag{6.1}$$

If attention is confined to the magnetic field initially, then the loss-less system reduces to an ideal inductance. The electrical energy supplied to an N-turn ideal coil at voltage v and current i in time δt follows from Faraday's law as

$$\delta W_e = vi\delta t = N\frac{d\Phi}{dt}i\delta t = Nid\Phi = Fd\Phi \tag{6.2}$$

where F is the m.m.f.

If no movement takes place, δW_m is zero and from Equation (6.1), $\delta W_e = \delta W_f$ and all the energy is stored in the magnetic field. In general, the change in magnetic stored field energy between coil-flux levels Φ_1, and Φ_2 is given by

$$W_{f2} - W_{f1} = \int_{\Phi_1}^{\Phi_2} Fd\Phi \tag{6.3}$$

which with a magnetically-linear system may be represented by the shaded area on a Φ–F diagram as indicated in Figure 6.2. In particular, if the current and flux increase from zero giving final values of

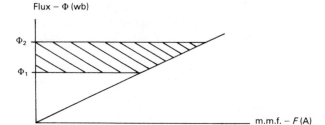

Figure 6.2 Stored magnetic field energy in magnetically-linear system

m.m.f. and flux of F and Φ respectively, then the stored field energy follows as

$$W_f = \int_0^\Phi F\,d\Phi = \tfrac{1}{2}F\Phi \tag{6.4}$$

But from the magnetic-circuit relationships developed in Section 4.4 and the definition of inductance, the stored field energy may be expressed in any of the following alternative forms:

$$W_f = \tfrac{1}{2}\mathscr{R}\Phi^2 = \tfrac{1}{2}\Lambda F^2 = \tfrac{1}{2}Li^2 \tag{6.5}$$

Each form is equally valid, and problem solving may be much simplified by suitable choice of the most convenient form in each particular case.

6.3 Mechanical work and force in a singly-excited magnetically-linear system

A *singly-excited* system is one possessing only one source of m.m.f. A simple example to consider is an idealised attracted-armature relay with a d.c. supply, as represented in Figure 6.3. The model assumes no flux leakage and infinitely permeable iron. Under initial conditions,

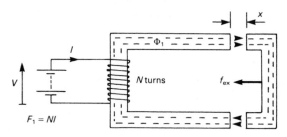

Figure 6.3 Idealised attracted-armature relay

the circuit contains a double airgap of length x, flux Φ_1, flows around the magnetic circuit, and the coil current sets up a steady m.m.f. F_1.

It is assumed that, under the influence of the attractive force f_{ex}, the armature moves a small distance δx towards the fixed member. With the reduced airgap, the flux increases to Φ_2, but the steady m.m.f. returns to F_1, because the current in the static d.c. circuit is unchanged. Since the system is doing work, δW_m is negative, and the energy balance equation may be rearranged as

$$-\delta W_m = \delta W_e - \delta W_f \tag{6.6}$$

But $\delta W_f = W_{f2} - W_{f1}$ where subscripts 1 and 2 refer to conditions before and after movement respectively. The work done therefore becomes

$$-\delta W_m = \delta W_e + W_{f1} - W_{f2} \tag{6.7}$$

The components of Equation (6.7) can be identified on the $\Phi - F$ operating diagram for the relay, shown in Figure 6.4. δW_e is the

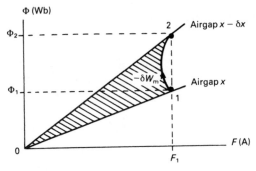

Figure 6.4 Φ–F diagram for operation of idealised relay

integral under the operating curve (of undefined shape) of area $\Phi_2 21 \Phi_1 \Phi_2$ on the figure. W_{f1} is $\frac{1}{2} F_1 \Phi_1$ or the area of triangle $0\Phi_1 10$. The sum of these two energy components is the total figure area $0\Phi_2 210$. Subtraction of W_{f2} or the triangle of area $0\Phi_2 20$ gives the work done $-\delta W_m$, which may be identified as the shaded area in the diagram. We may therefore say that

The work done during movement is given by the area bounded by the two operating lines and the locus of the movement in the operating point.

It is not generally known how the locus varies during movement, but two ideal limiting conditions of relay operation may be identified. These being:

(a) *very slow operation*, in which induced e.m.f. is assumed negligible, and so current stays constant. This is *constant-F operation*.
(b) *very fast operation*, in which Φ has no time to vary (since infinite induced e.m.f. would be implied). This is *constant-Φ operation*.

These two operating conditions are illustrated in Figures 6.5(a) and (b) respectively.

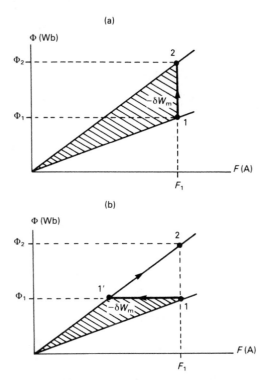

Figure 6.5 Limiting conditions of idealised relay operation   (a) constant F (b) constant Φ

For constant-F operation, it can be seen from Figure 6.5(a) that:

$$\delta W_e = F_1(\Phi_2 - \Phi_1) \tag{6.8}$$

$$-\delta W_m = \tfrac{1}{2}F_1(\Phi_2 - \Phi_1) \tag{6.9}$$

and

$$\delta W_f = -\delta W_m = \tfrac{1}{2}\delta W_e \tag{6.10}$$

Hence, for a *magnetically-linear* electromagnetic system operating at *constant m.m.f.*, where movement takes place, electrical energy is drawn from the supply, *half* of which is converted to mechanical work (energy), and *half* of which is stored in the magnetic field. This is sometimes referred to as the *50–50 rule of electromagnetics*.

With f_{ex} the force developed by the system in the x direction, and δW_m the mechanical energy input to the system, then

$$\delta W_m = -f_{ex}\delta x = -\delta W_f \tag{6.11}$$

and

$$f_{ex} = \frac{\partial W_f}{\partial x}\bigg|_{F \text{ constant}} \tag{6.12}$$

The partial differential form must be used here since W_f is a function of both F and Φ.

Constant-Φ operation must be considered in two stages. The instantaneous movement phase occurs first, when the operating point moves at constant flux from 1 to 1', no electrical energy transfer takes place, and mechanical work is done by extraction of field energy. In this phase

$$\delta W_m = -f_{ex}\,\delta x = \delta W_f \tag{6.13}$$

and

$$f_{ex} = -\frac{\partial W_f}{\partial x}\bigg|_{\Phi \text{ const}} \tag{6.14}$$

also, from the geometry of Figure 6.5(a), it can be deduced that

$$-\delta W_m = \tfrac{1}{2}F_1(\Phi_2 - \Phi_1)\Phi_1/\Phi_2 \tag{6.15}$$

In the second, or recovery, phase of constant-Φ operation, the operating point moves from 1' to 2 up the final operating line. The movement has already finished, and so no further work is done. Electrical energy is drawn from the supply and stored in the field. From the diagram, this component of energy follows as

$$\delta W_e = F_1(\Phi_2^2 - \Phi_1^2)/(2\Phi_2) \tag{6.16}$$

It should be emphasised that the mechanical work done and the energy transfers are only calculable when the locus of the movement is known. However, the force equations

$$f_{ex} = \frac{\partial W_f}{\partial x}\bigg|_{F \text{ const}} = -\frac{\partial W_f}{\partial x}\bigg|_{\Phi \text{ const}} \tag{6.17}$$

are generally applicable to linear systems, irrespective of how the movement takes place. This, perhaps surprising fact may be appreciated diagramatically if the limiting condition of infinitesimally small movement is considered under constant-F and constant-Φ operation. The effect of this on Figures 6.5(a) and (b) is to bring the operating lines, and hence the operating points 1 and 2 and the fluxes Φ_1 and Φ_2 into close proximity. It should then be obvious that, in this limiting condition, the areas representing mechanical work in the two diagrams are equal.

Alternatively, the two forms of force expression in Equation (6.17) may be shown to be equal mathematically, using the magnetic-circuit formulae of Section 4.4. Since

$$W_f = \tfrac{1}{2}F^2 \Lambda$$

$$\frac{\partial W_f}{\partial x}\bigg|_{F \text{ const}} = \tfrac{1}{2}F^2\frac{d\Lambda}{dx} \tag{6.18}$$

but $\Lambda = 1/\mathcal{R}$ and $\dfrac{d\Lambda}{dx} = \dfrac{d\Lambda}{d\mathcal{R}}\dfrac{d\mathcal{R}}{dx} = \dfrac{-1}{\mathcal{R}^2}\dfrac{d\mathcal{R}}{dx}$

Hence from Equation (6.18)

$$\frac{\partial W_f}{\partial x}\bigg|_{F \text{ const}} = -\frac{1}{2}\frac{F^2}{\mathcal{R}^2}\frac{d\mathcal{R}}{dx} = -\frac{1}{2}\Phi^2\frac{d\mathcal{R}}{dx} \tag{6.19}$$

Then using the alternative field energy expression in terms of

$$W_f = \frac{1}{2}\Phi^2\mathcal{R}$$

gives

$$-\frac{\partial W_f}{\partial x}\bigg|_{\Phi \text{ const}} = -\frac{1}{2}\Phi^2\frac{d\mathcal{R}}{dx} \tag{6.20}$$

Equations (6.19) and (6.20) are evidently equal, demonstrating the equal validity of the two force expressions. A typical application of the formulae is the calculation of the alignment force between overlapping magnetised iron surfaces. The arrangement is illustrated in Figure 6.6(a), where an N-turn coil carrying current i drives flux across an airgap of length x. The standard assumptions are made of zero magnetic leakage and fringing, in addition to the necessary linearity condition that all the iron parts of the magnetic circuit have zero reluctance. A vertical section through the arrangement is shown in Figure 6.6(b), where the overlap is shown as y. Then, from Equation (6.18).

(a)

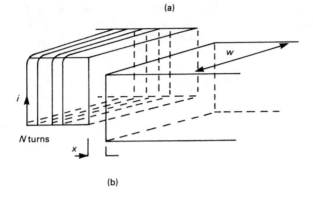

(b)

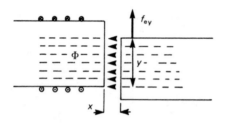

Figure 6.6 Force of alignment between magnetised iron surfaces   (a) physical
arrangement   (b) vertical section

$$f_{ey} = \left.\frac{\partial W_f}{\partial y}\right|_{F\,\text{const}} = \tfrac{1}{2}F^2\frac{d\Lambda}{dy} \tag{6.21}$$

But $F = Ni$ and $\Lambda = \mu_o\,yw/x$

Hence
$$f_{ey} = \tfrac{1}{2}N^2 i^2 \mu_o w/x \tag{6.22}$$

This is a force of constant magnitude in the positive direction, which
is to increase the overlap y.

Note that this formula must be applied with care, since the force
disappears when the surfaces are fully aligned; a fact not self-evident
from the equation.

6.4 Comparative force levels in electromagnetic and electrostatic systems

Since equations in electromagnetic and electrostatic systems are analogous, complementary force equations may be developed for each

system to allow force levels to be compared. Figure 6.7(a) shows the electromagnetic arrangement of N and S polepieces of area A separated by a distance x. If the airgap flux density is $\mathbf{B}$ T, then from Equation (6.5)

$$W_f = \tfrac{1}{2}\mathcal{R}(\mathbf{B}A)^2 = \tfrac{1}{2}x\mathbf{B}^2 A/\mu_o \tag{6.23}$$

then

$$f_{ex} = \frac{-\partial W_f}{\partial x}\bigg|_{\Phi\ \text{const}} = -\mathbf{B}^2 A/(2\mu_o) \tag{6.24}$$

where the minus sign indicates an attractive force.

Equation (6.24) may be expressed in terms of force/area or pressure as

$$f_{ex}/A = -\mathbf{B}^2/(2\mu_o)\ N/m^2 \tag{6.25}$$

With circuits containing iron, saturation limits $\mathbf{B}$ to typically 1.5 T. This gives a force/area or pressure value of about 9×10^5 N/m^2.

A corresponding electrostatic force of attraction occurs between the parallel plates of a charged capacitor, as shown in Figure 6.7(b). The electrostatic quantity corresponding to $\mathbf{B}$ is the electric charge density or flux density Q/A. Hence the electrostatic force/area or pressure expression follows by analogy as

$$f_{ex}/A = Q^2/(2\varepsilon_o A^2) \tag{6.26}$$

assuming the relative permittivity of air to be 1. In the electrostatic case, the force level is limited by the voltage gradient (V/x) in the air between the plates. From the capacitance Equations (2.24) and (2.25), Equation (6.25) may be rewritten in terms of voltage and gradient as

$$f_{ex}/A = -\tfrac{1}{2}\varepsilon_o\left(\frac{V}{x}\right)^2 \tag{6.27}$$

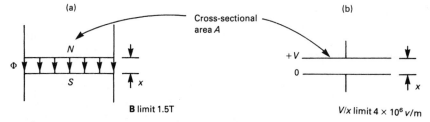

Figure 6.7 Comparative force levels in electromagnetic and electrostatic systems   (a) electromagnetic   (b) electrostatic

A typical limiting value for V/x in air at an x of 1 mm is 4×10^6 V/m. Values in excess of this would cause electrical discharge and damage to the plates. At this limiting value, f_{ex}/A is about 70 N/m^2, or well over ten thousand times smaller than the limiting electromagnetic force. The V/x limit may be increased by placing the plates in a vacuum, but at obvious increased cost. The electrostatic device has the added disadvantage that very high voltages must be employed to obtain any significant force.

This enormous difference in force levels is the fundamental reason why electromagnetic devices are almost always used in preference to electrostatic ones for electromechanical energy conversion.

6.5 Doubly-excited systems

Self-evidently, a *doubly-excited electromagnetic system* is one containing two sources of m.m.f. and this is the normal situation occurring in the majority of rotating electrical machines, as well as in some actuators and electromechanical transducers. If the m.m.f.s are set up by two coils, 1 and 2, with magnetic coupling between them, then

$$W_f = \tfrac{1}{2}F_1\Phi_1 + \tfrac{1}{2}F_2\Phi_2$$
$$= \tfrac{1}{2}N_1I_1\Phi_1 + \tfrac{1}{2}N_2I_2\Phi_2 \tag{6.28}$$

But from the theory developed in Section 3.3

$$\Phi_1 = \Phi_{11} + \Phi_{12}$$

and

$$\Phi_2 = \Phi_{22} + \Phi_{21}$$

where Φ_{11} is the self flux set up by I_1 in coil 1 and Φ_{12} is the mutual flux set up in coil 1 by current I_2 in coil 2. The components of Φ_2 follow correspondingly. Then by expanding Equation (6.28) in terms of flux components, and applying the self and mutual inductance definitions (Equations (3.13) and (3.14) in Section 3.3), the stored field energy expression may be reduced to

$$W_f = \tfrac{1}{2}L_{11}I_1^2 + \tfrac{1}{2}L_{22}I_2^2 + I_1I_2M \tag{6.29}$$

The straight-line force in the x direction then follows directly as

$$f_{ex} = \frac{\partial W_f}{\partial x}\bigg|_{F \text{ const}} = \tfrac{1}{2}I_1^2\frac{dL_{11}}{dx} + \tfrac{1}{2}I_2^2\frac{dL_{22}}{dx} + I_1I_2\frac{dM}{dx} \tag{6.30}$$

Two distinct force mechanisms may be identified in the above expression. The first two terms contain only one current, and the force is derived from change in self-inductance, or self-flux linkage, with position. Forces of this type are the only ones present in singly-excited systems, and are termed *alignment* or *attraction forces* because of their manner of operation. The presence of the current-squared term indicates that the direction of these forces is independent of current polarity. In contrast, the third term contains both currents, and depends on the change in mutual inductance with position. Since this force component is developed via the interaction between the two currents or m.m.f.s, it is absent in singly-excited systems, and it is termed a *force of interaction*. Evidently, change in polarity here of one current reverses the direction of the force.

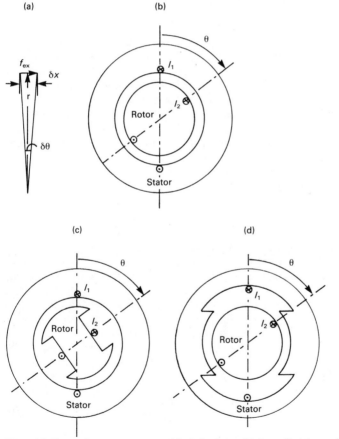

Figure 6.8 Torque in rotary systems   (a) derivation   (b) 'round' stator and rotor
(c) 'round' stator salient rotor   (d) salient stator 'round' rotor

Interaction forces may be calculated using the familiar $\mathbf{B}Il$ formula, where I represents one current, and $\mathbf{B}$ the flux density set up by the other current. Attraction or alignment forces are best calculated via the energy equation above.

The torque developed in a rotary system may be similarly derived. By reference to Figure 6.8(a), if force f_{ex} operates at a radius r in a rotary system, the torque T developed is given by

$$T = f_{ex}r$$

but since $\delta x = r\delta\theta$, then in the limit as $\delta x \to 0$

$$T = f_{ex}\frac{dx}{d\theta} = \frac{\partial W_f}{\partial \theta}\bigg|_F \text{ const} \tag{6.31}$$

which follows from Equation (6.30). Hence from Equation (6.29)

$$T = \tfrac{1}{2}I_1^2\frac{dL_{11}}{d\theta} + \tfrac{1}{2}I_2^2\frac{dL_{22}}{d\theta} + I_1 I_2 \frac{dM}{d\theta} \tag{6.32}$$

As before, the first two terms result from alignment or attraction forces, and the third from an interaction force.

The addition of the second current-carrying coil or source of m.m.f. in the double-excited rotational system introduces the interaction force and its consequent torque component. It is this component which is responsible for electromechanical energy conversion in the vast majority of rotating electrical machines. A primitive machine is outlined in Figure 6.8(b) in which only mutual inductance varies with position, and hence only interaction force contributes to the torque. Such a machine is said to have a *round* stator and *round* rotor. Often however, manufacturing techniques dictate that the rotor may not be treated as uniformly round. Instead, it has significant shaping or *saliency*, as shown in Figure 6.8(c). This results in stator-current-driven flux being dependent on rotor position. In turn this makes stator self-inductance variable, and introduces alignment or attraction force. The corresponding torque component is called a *reluctance* torque, and reluctance motors which exploit the effect are produced with salient rotors containing no windings. A device with salient stator and round rotor is shown in Figure 6.8(d). Here the rotor-current-driven flux and rotor self-inductance are evidently position dependent, and a corresponding rotor saliency torque is produced.

If idealising assumptions are made about inductance variations with position, torque can be calculated using Equation (6.32).

6.6 Non-linear systems

Magnetic non-linearity may be introduced by considering the effect

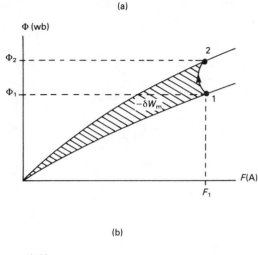

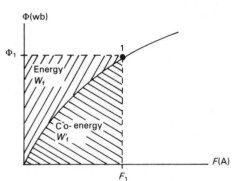

Figure 6.9 Non-linear Φ–F diagrams for attracted-armature relay   (a) operating characteristics   (b) energy and co-energy

of saturation on the attracted-armature relay of Section 6.3. Figure 6.9(a) shows the modified Φ–F operating characteristics. The work done during movement is evidently represented by the corresponding area in the diagram, but it is no longer a simple triangle, and hence is generally more difficult to calculate.

Before considering constant m.m.f. (very slow) and constant Φ (very fast) operation, the quantity *co-energy* must be defined. From Equation (6.4), the stored field energy W_f at operating point 1 is given by

$$W_f = \int_0^{\Phi_1} F\mathrm{d}\Phi \tag{6.33}$$

The corresponding field co-energy W'_f is defined as

$$W'_f = \int_o^{F_1} \Phi dF \tag{6.34}$$

Areas corresponding to W_f and W'_f are indicated in Figure 6.9(b). Evidently in the exactly linear case field energy and co-energy are equal. Magnetic saturation has the effect of making the co-energy exceed the energy.

The limiting operating conditions are shown in Figure 6.10(a) for constant m.m.f. and 6.10(b) for constant Φ. For constant m.m.f.

$$-\delta W_m = \delta W'_f \text{ (since co-energy is increasing)} \tag{6.35}$$

(a)

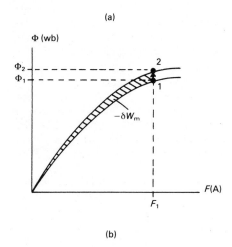

(b)

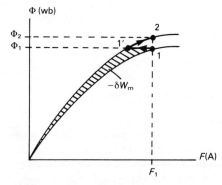

Figure 6.10 Limiting conditions of non-linear relay operation   (a) very slow – constant F (b) very fast – constant Φ

and for constant flux

$$-\delta W_m = -\delta W_f \text{ (since energy is decreasing)} \tag{6.36}$$

As for the linear system, with infinitesimally small movement the work done tends to the same value for the two limiting conditions. Hence the force developed by the relay is given by the two alternative forms

$$f_{ex} = \left.\frac{\partial W_f'}{\partial x}\right|_{F\ const} = -\left.\frac{\partial W_f}{\partial x}\right|_{\Phi\ const} \tag{6.37}$$

and corresponding expressions may be developed for torque in rotational systems.

The effect of saturation is to introduce some field energy storage into the iron. However, if Φ is constant, the m.m.f. and energy stored in the iron depend only on the dimensions of the iron circuit, and are independent of the airgap length x. Hence in the constant Φ version of Equation (6.37), rate of change of airgap energy alone need be considered. Force and torque equations developed for the linear case remain valid therefore in the non-linear case, provided that airgap values, rather than overall circuit values, are used. Consequently, Equation (6.22) for overlap force applies, with the proviso that the m.m.f. Ni is the airgap value. Equation (6.24) for attraction force remains valid without qualification, since **B** is the flux density at the airgap.

Equations (6.37) apply also to systems with multiple excitation, but their evaluation in these cases may become exceedingly complex.

6.7 Bibliography

MASON J. C., *BASIC Numerical Mathematics*, Butterworths, London (1983)

WORKED EXAMPLES

Example 6.1: LOCUS: dynamic performances of d.c. relay

The theory of an idealised relay is given in Section 6.3.

Considerations of the locus of movement are confined there to two limiting cases. If friction is neglected and the mass of the moving member is known, however, it is possible to estimate the actual locus and time taken for movement in individual cases by using a simple time-stepping approach. Develop a program to do this, which displays all the relevant data at successive time instants.

```
10 REM       LOCUS - Dynamic performance of "real" d.c.
20 REM       relay, neglecting magnetic saturation,
25 REM       flux leakage, and fringing
30 P1 = 3.14159265
40 M0 = 4*P1*1.E-7
50 PRINT
60 PRINT
70 PRINT "Dynamic performance of magnetically-linear";
75 PRINT " d.c. relay"
80 PRINT "--------------------------------------------------";
85 PRINT "------------"
90 PRINT
100 PRINT "mass of moving member (gm) ";
110 INPUT M
120 M = M/1000
130 PRINT
140 PRINT "cross-sectional-area of ONE airgap (mm^2) ";
150 INPUT A
160 A = A*1E-6
170 PRINT
180 PRINT "number of turns on coil ";
190 INPUT N
200 K1 = N^2*M0*A/(4*M)
210 K2 = 2/(N^2*M0*A)
220 K3 = K1*M
230 K4 = 0.5*N*M0
240 PRINT
250 PRINT "d.c. supply voltage ";
260 INPUT V
270 PRINT
280 PRINT "total circuit resistance (ohms) ";
290 INPUT R
300 PRINT
310 PRINT "initial airgap - OPEN (mm) ";
320 INPUT X0
330 X0 = X0/1000
340 PRINT
350 PRINT "final airgap - CLOSED (mm) ";
360 INPUT X4
370 X4 = X4/1000
380 IF X4>X0 THEN GOTO 800
390 IF K4*V/(R*X4)>=1.0 THEN GOTO 830
400 PRINT
410 PRINT "time-step interval (microS) ";
420 INPUT T1
430 T1 = T1*1E-6
440 PRINT
450 PRINT "max time for movement allowed (microS) ";
460 INPUT T2
470 T2 = T2*1E-6
480 PRINT
490 PRINT "time interval for output (microS) ";
500 INPUT T3
510 T3 = T3*1E-6
520 N2 = T3/T1
530 PRINT
540 PRINT
550 PRINT " t(ms)    x(mm)   dx/dt(m/s)   i(A)       f(N)";
555 PRINT "      B(T)"
560 PRINT "---------------------------------------------";
```

```
565 PRINT "----------------"
570 PRINT
580 T = 0
590 N1 = 1
600 I = V/R
610 X = X0
620 X1 = 0
630 V1 = I/X
640 V2 = V1^2
650 F = -K3*V2
660 X2 = -K1*V2
670 B = K4*V1
680 IF N1=1 THEN PRINT T*1000; TAB(10);X*1000; TAB(20);X1;
690 IF N1=1 THEN PRINT TAB(30); I;TAB(40);F;TAB(50);B
700 T = T + T1
710 IF T>T2 THEN GOTO 870
720 X1 = X1 + X2*T1
730 X = X + X1*T1
740 IF X<=X4 THEN GOTO 910
750 I1 = (V - R*I)*K2*X + X1*V1
760 I = I + I1*T1
770 N1 = N1 + 1
780 IF N1>N2 THEN N1 = 1
790 GOTO 630
800 PRINT "Final airgap must be SMALLER";
805 PRINT " than initial airgap"
810 PRINT "Please enter data again"
820 GOTO 300
830 PRINT
840 PRINT "static flux density in closed position > 1T";
845 PRINT " - too high"
850 PRINT "Please increase circuit resistance or reduce";
855 PRINT " turns or voltage"
860 GOTO 170
870 PRINT
880 PRINT "time interval exceeded before movement"
890 PRINT "completed - increase time interval allowed"
900 GOTO 440
910 STOP
920 END


Dynamic performance of magnetically-linear d.c. relay
-----------------------------------------------------

mass of moving member (gm) ?1

cross-sectional-area of ONE airgap (mm^2) ?25

number of turns on coil ?1000

d.c. supply voltage ?12

total circuit resistance (ohms) ?8

initial airgap - OPEN (mm) ?2

final airgap - CLOSED (mm) ?1

time-step interval (microS) ?.5
```

```
max time for movement allowed (microS) ?1000
```

```
time interval for output (microS) ?50
```

t(ms)	x(mm)	dx/dt(m/s)	i(A)	f(N)	B(T)
0	2	0	1.5	-4.418	0.4712
5E-2	1.994	-0.2209	1.496	-4.418	0.4713
0.1	1.978	-0.4419	1.484	-4.421	0.4714
0.15	1.95	-0.6631	1.464	-4.429	0.4718
0.2	1.911	-0.8848	1.438	-4.443	0.4726
0.25	1.862	-1.108	1.404	-4.467	0.4738
0.3	1.8	-1.332	1.363	-4.502	0.4757
0.35	1.728	-1.558	1.315	-4.55	0.4783
0.4	1.645	-1.787	1.261	-4.615	0.4816
0.45	1.549	-2.02	1.198	-4.698	0.4859
0.5	1.442	-2.257	1.128	-4.802	0.4913
0.55	1.323	-2.5	1.049	-4.932	0.4979
0.6	1.192	-2.751	0.9596	-5.09	0.5058
0.65	1.048	-3.01	0.8594	-5.281	0.5152

```
STOP at line 910
```

```
Dynamic performance of magnetically-linear d.c. relay
------------------------------------------------------------
```

```
mass of moving member (gm) ?50000
```

```
cross-sectional-area of ONE airgap (mm^2) ?25
```

```
number of turns on coil ?1000
```

```
d.c. supply voltage ?12
```

```
total circuit resistance (ohms) ?8
```

```
initial airgap - OPEN (mm) ?2
```

```
final airgap - CLOSED (mm) ?1
```

```
time-step interval (microS) ?100
```

```
max time for movement allowed (microS) ?150000
```

```
time interval for output (microS) ?10000
```

t (ms)	x (mm)	dx/dt (m/s)	i (A)	f (N)	B(T)
0	2	0	1.5	-4.418	0.4712
10	1.996	-8.846E-4	1.499	-4.434	0.4721
20	1.982	-1.776E-3	1.499	-4.49	0.4751
30	1.96	-2.683E-3	1.498	-4.588	0.4803
40	1.928	-3.615E-3	1.497	-4.735	0.4878
50	1.887	-4.58E-3	1.496	-4.937	0.4981
60	1.836	-5.593E-3	1.495	-5.206	0.5116
70	1.775	-6.668E-3	1.494	-5.562	0.5288
80	1.703	-7.825E-3	1.492	-6.032	0.5507
90	1.618	-9.09E-3	1.49	-6.66	0.5786
100	1.52	-1.05E-2	1.487	-7.515	0.6146
110	1.407	-1.212E-2	1.483	-8.72	0.662
120	1.277	-1.403E-2	1.476	-10.5	0.7265
130	1.125	-1.638E-2	1.465	-13.32	0.8182

STOP at line 910

Program notes
(1) The solution is based on the solution of two equations; one electromechanical and the other electromagnetic. With reference to Figure 6.3, if A is the cross-sectional area of the relay iron, then neglecting flux leakage, fringing and magnetic saturation, the force f_{ex} follows from Equation (6.18). Using the dot notation to indicate differentiation with respect to time, then the acceleration $\ddot{x}$ of the moving member of mass m when the coil carries general current i follows as

$$\ddot{x} = -\frac{N^2 i^2 \mu_o A}{4x^2 m} \tag{6.38}$$

The electromagnetic equation gives the supply voltage V as the sum of the ohmic voltage drop and the induced e.m.f. or

$$V = Ri + N\frac{d}{dt}\left(\frac{Ni\mu_o A}{2x}\right) \tag{6.39}$$

Since both i and x vary with time, performing the differentiation gives

$$V = Ri + \frac{N^2 \mu_o A}{2x^2}(x\dot{i} - i\dot{x})$$

which may be re-arranged to give

$$\dot{i} = (V - Ri)\frac{2x}{N^2\mu_o A} + \frac{i\dot{x}}{x} \tag{6.40}$$

If a small time increment δt is made at each step, then the new speed value $\dot{x}_{n+1}$ may be approximated from the previous value $\dot{x}_n$ using the formula

$$\dot{x}_{n+1} = \dot{x}_n + \ddot{x}_n \, \delta t \tag{6.41}$$

and similarly

$$x_{n+1} = x_n + \dot{x}_n \, \delta t \tag{6.42}$$

$$i_{n+1} = i_n + \dot{i}_n \, \delta t \tag{6.43}$$

(2) Constants K1 to K4 are used to represent the parts of Equations (6.38) and (6.40) which do not vary with time (lines 200–230).

(3) Data validation is included to check that the final airgap is smaller than the initial airgap, and that the steady flux density occurring in the closed position does not exceed 1T. The second test is to ensure that the assumption of magnetically-unsaturated iron is reasonable.

(4) Obviously the accuracy of the program must depend on the time-step length used (T1). Since the user may not wish to view the results at each time step, a second time variable (T3) determines the interval at which the results are to be printed. Hence every 'N2th' result is printed, where $N2 = T3/T1$ (line 520). A time limit (T2) is also input, to prevent the program from running indefinitely in the event of an error in the input data.

(5) The initial values of the variables are input over lines 580–620, these being time (zero), step number (1), steady-state current (V/R), x (open position), and velocity (zero).

(6) Lines 630–790 cover the time stepping. Auxiliary variables V1 and V2 are calculated at each step since they are used several times. Force is calculated at line 650 and $\ddot{x}$ (X2) using Equation (6.38) appears at line 660. Equation (6.40) for $\dot{i}$ (I1) is at line 750, and the three integration Equations (6.41) to (6.43) appear at lines 720, 730, and 760 respectively.

(7) Output is printed whenever N1 has the value 1. Hence the initial conditions are printed. Then at each subsequent time step, N1 is increased by one. A test is made to establish whether N1 exceeds N2, which, if true, results in N1 being reset to 1 once more. This ensures output of results at regular intervals of T3 (i.e. N2 time intervals).

(8) If the time interval is not exceeded (see line 710) the run is completed when x reaches the specified closed position value (line 740).

(9) The first run approaches the very fast (or constant Φ) condition, when the moving member has a small mass. Note that this also implies a constant force value during movement. In the second run, the member is assumed attached to a large mass, and the constant m.m.f. condition is approximated. Loci for the two runs, with times, are plotted on a Φ–F diagram in Figure 6.11. (The end timings were obtained by repeating the runs with T3 values reduced from those used for the listed results.) Note that $\dot{x}$ is not zero when the closed

WMETER: calculation of scale graduations on an electrodynamic wattmeter

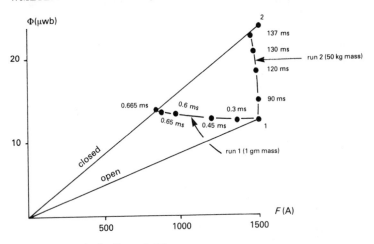

Figure 6.11 Results for Example 6.1

position is reached, implying some form of physical restraint or end-stop preventing further movement.

(10) An approximate technique of this type requires some check on accuracy. The simplest to apply is to repeat runs with successively smaller T1 values but the same T3 value until a consistent result is obtained. The user may check that the runs above satisfy this criterion.

Example 6.2: WMETER: calculation of scale graduations on an electrodynamic wattmeter

An electrodynamic wattmeter consists essentially of a fixed circular current coil and a pivoted voltage coil in air. The mutual inductance between the coils, when aligned, is 0.3 mH. The voltage-coil circuit includes a resistance of 10 kΩ which swamps the inductive reactance of the voltage coil at the designed operating frequency of 50 Hz. Hence the voltage-coil current is proportional to, and in phase with, the voltage across it. Self-inductances are constant, and the mutual inductance variation may be assumed cosinusoidal with the angle θ between the coils. For zero power, the coils are orthogonal, and rotation of the voltage coil is restrained by a spring having a linear relationship between restoring torque and angle. The spring produces zero torque with orthogonal coils, and peak restoring torque when the coils are aligned.

Write a program to facilitate selection of a suitable spring to make the scale reasonably linear. The meter is designed to operate at powers up to 1 kW.

```
10 REM      WMETER - Scale graduations for electrodynamic
20 REM      wattmeter. Power-angle equation solved
25 REM      iteratively using Newton's Method
30 REM      ( x(n+1) = x(n) - f(x(n))/f'(x(n)) )
40 DATA 0.3E-3, 10000, 1E-4, 3.14159265
50 READ M, R, E, P1
60 PRINT
70 PRINT "Scale graduations for electrodynamic wattmeter"
80 PRINT "-----------------------------------------------"
90 PRINT
100 PRINT "(M between aligned coils = "; M;
105 PRINT "H,   V-coil R = "; R; " Ohms)"
110 PRINT
120 PRINT "Spring-restoring torque ";
125 PRINT "when coils aligned (Nm) ";
130 INPUT T1
140 K1 = 2*T1/P1
150 PRINT
160 PRINT "Power (W)      Angle (deg)"
170 PRINT "---------      -----------"
180 PRINT
190 PRINT TAB(5); "0"; TAB(14); "90"
200 FOR P = 100 TO 1000 STEP 100
210   T0 = P1/4
220   K2 = P*M/R
230   D = FNA(T0)
240   T0 = T0 - D
250   IF ABS(D/T0)>E THEN GOTO 230
260   T0 = T0*180/P1
270   PRINT TAB(3); P; TAB(14); T0
280   NEXT P
290 PRINT
300 PRINT "Another spring (Y/N) ";
310 INPUT Q$
320 IF Q$ = "Y" THEN GOTO 60
330 STOP
340 REM *** Definition of function FNA(x) = f(x)/f'(x) ***
350 DEF FNA(X) = (K1*X + K2*SIN(X) - T1)/(K1 + K2*COS(X))
360 END
```

```
Scale graduations for electrodynamic wattmeter
-----------------------------------------------

(M between aligned coils = 3E-4H,   V-coil R = 10000 Ohms)

Spring-restoring torque when coils aligned (Nm) ?2E-5
```

WMETER: calculation of scale graduations on an electrodynamic wattmeter

```
Power  (W)        Angle  (deg)
----------        ------------

      0           90
    100           76.8537962
    200           65.4423283
    300           56.3042079
    400           49.1519019
    500           43.5194487
    600           39.0119359
    700           35.3394135
    800           32.2961232
    900           29.7358141
   1000           27.5530633

Another spring  (Y/N)  ?Y

Scale graduations for electrodynamic wattmeter
----------------------------------------------------

(M between aligned coils = 3E-4H,   V-coil R = 10000 Ohms)

Spring-restoring torque when coils aligned (Nm) ?1E-4

Power  (W)        Angle  (deg)
----------        ------------

      0           90
    100           87.3029907
    200           84.6237551
    300           81.9792376
    400           79.3848248
    500           76.8537961
    600           74.3969949
    700           72.0227175
    800           69.7367912
    900           67.5427889
   1000           65.4423283

Another spring  (Y/N)  ?N

STOP at line 330
```

Program notes
(1) With instantaneous currents i_1 and i_2 passing through the fixed (current) coil and moving (voltage) coil, the instantaneous torque T acting on the moving coil follows from Equation (6.32) as

$$T = -i_1 i_2 M \sin \theta \qquad (6.44)$$

The inertia of the instrument prevents it from responding to the rapid time variation of the instantaneous torque, and hence it responds to the mean value of torque produced. If I and V are the r.m.s. magnitudes of the current and voltage (assumed sinusoidal) in the circuit whose power is being measured, then from the theory of Section 3.5, if the voltage coil swamp resistance is R, the mean torque on the coil $\bar{T}$ follows as

$$\bar{T} = - VI \cos \varphi M \sin \theta = - \bar{P} M \sin \theta \qquad (6.45)$$

where $\bar{P}$ is the mean power to be measured.

If the restoring torque produced by the spring is T_1 when $\theta = 0$ and zero when $\theta = \pi/2$, its torque equation must be

$$T = T_1 (1 - 2\theta/\pi) \qquad (6.46)$$

Since deflection and restoring torques act in opposition, the $\bar{P} - \theta$ relationship follows from the sum of Equations (6.45) and (6.46) as

$$(2T_1/\pi)\theta + (\bar{P}M/R) \sin \theta - T_1 = 0 \qquad (6.47)$$

Transcendental Equation (6.47) may be solved for θ in steps of $\bar{P}$ up to 1 kW using Newton's method. (The reader unfamiliar with Newton's Method is referred to Mason (1983). The algorithm is stated in a REM statement at line 30 in the program.)

(2) A DATA statement is used to preset values of M, R, the maximum error allowed in the Newton's method iteration (E), and π. The peak restoring torque of the spring (T1), is input as data to allow its value to be varied.

(3) Values of θ for powers from 100 W to 1 kW in steps of 100 W are calculated in a loop covering lines 200–280. θ is initially set to $\pi/4$ to start the iterative process used in Newton's Method. The method requires use of Equation (6.47) in function form, divided by its differential with respect to θ. This is obtained via the function FNA defined at line 350. The method is actually implemented from lines 230–250. When the proportional change produced in the value of θ (TO) by the iterative process is less than E, the angle is converted to degrees and displayed.

(4) The two runs show the effect of variation of spring stiffness on the scale. The weaker spring with T1 of 20 μNm gives a reasonable scale spread of about 60°, but the graduations are cramped at the top end. The stronger spring with T1 of 100 μNm gives much more even graduations, but with the whole scale spread confined within about 25°. Figure 6.12 shows how the two scales would look.

(5) The theory assumed the voltage and current to be sinusoidal. In fact, this restriction is unnecessary, and the meter gives a true reading of mean power, whatever the shape of voltage or current waveforms. The reader should give some thought as to the reason for this.

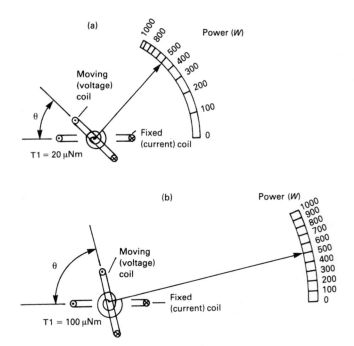

Figure 6.12 Resulting wattmeter scales for Example 6.2

Example 6.3: TORQUE: torque-angle characteristics for basic rotary machine

The method of calculating torque in a doubly-excited rotary machine from products of currents and rates of change of inductance with position is indicated in Section 6.5. Making assumptions that these inductance variations are cosinusoidal or zero where appropriate, develop a program to display torque against rotor angle for particular current values. Additional input data required includes saliency information for the rotor and stator, corresponding self-inductance values, and a coefficient of coupling value for the stator and rotor coils when aligned.

```
10 REM      TORQUE - Torque vs angle characteristics for
15 REM              rotary machine
20  P1 = 3.14159265
30 PRINT
40 PRINT
50 PRINT "Torque-angle characteristics for rotary ";
55 PRINT "machine"
```

```
 60 PRINT "--------------------------------------------";
 65 PRINT "--------"
 70 PRINT "(assuming cosinusoidal inductance variation ";
 75 PRINT " with angle)"
 80 PRINT
 90 PRINT
100 PRINT "Instantaneous stator current (A) ";
110 INPUT I1
120 PRINT
130 PRINT "Instantaneous rotor current (A) ";
140 INPUT I2
150 PRINT
160 PRINT "Stator self-inductance when stator"
170 PRINT "and rotor coils aligned (mH) ";
180 INPUT L1
190 L1 = L1*1E-3
200 PRINT
210 PRINT "Is rotor Salient or Round (S/R) ";
220 INPUT R$
230 PRINT
240 IF R$="S" THEN PRINT "Stator self-inductance ";
245 IF R$="S" THEN PRINT "when stator"
250 IF R$="S" THEN PRINT "and rotor coils ";
255 IF R$="S" THEN PRINT "orthogonal (mH) ";
260 IF R$="S" THEN INPUT L0
270 IF R$="S" THEN L0 = L0*1E-3
280 PRINT
290 PRINT "Rotor self-inductance when rotor"
300 PRINT "and stator coils aligned (mH) ";
310 INPUT L2
320 L2 = L2*1E-3
330 PRINT
340 PRINT "Is stator Salient or Round (S/R) ";
350 INPUT S$
360 IF S$="S" THEN PRINT
370 IF S$="S" THEN PRINT "Rotor self-inductance ";
375 IF S$="S" THEN PRINT "when rotor"
380 IF S$="S" THEN PRINT "and stator coils ";
385 IF S$="S" THEN PRINT "orthogonal (mH) ";
390 IF S$="S" THEN INPUT L3
400 IF S$="S" THEN L3 = L3*1E-3
410 PRINT
420 PRINT "Coefficient of coupling between coils ";
425 PRINT "when aligned ";
430 INPUT K1
440 IF K1>1 THEN GOTO 650
450 IF K1<=0 THEN GOTO 650
460 M = K1*SQR(L1*L2)
470 PRINT
480 PRINT
490 PRINT "Angle (deg)        Torque (Nm)"
500 PRINT "-----------        -----------"
510 PRINT
520 FOR N = 0 TO 180 STEP 15
530   A = N*PI/180
540   T = -I1*I2*M*SIN(A)
550   IF R$="S" THEN T = T - 0.5*I1^2*(L1 - L0)*SIN(2*A)
```

```
560    IF S$="S" THEN T = T - 0.5*I2^2*(L2 - L3)*SIN(2*A)
570    PRINT TAB(4); N; TAB(16); T
580    NEXT N
590 PRINT
600 PRINT
610 PRINT "Another run (Y/N) ";
620 INPUT Q$
630 IF Q$="Y" THEN GOTO 30
640 STOP
650 PRINT
660 PRINT "Coupling coefficient must be +ve and <= 1"
670 PRINT "Please enter correct value"
680 GOTO 410
690 END
```

>RUN

Torque-angle characteristics for rotary machine

(assuming cosinusoidal inductance variation with angle)

Instantaneous stator current (A) ?10

Instantaneous rotor current (A) ?10

Stator self-inductance when stator
and rotor coils aligned (mH) ?50

Is rotor Salient or Round (S/R) ?R

Rotor self-inductance when rotor
and stator coils aligned (mH) ?30

Is stator Salient or Round (S/R) ?R

Coefficient of coupling between coils when aligned ?.9

Angle (deg)	Torque (Nm)
0	0
15	-0.90216
30	-1.7428
45	-2.4648
60	-3.0187
75	-3.3669
90	-3.4857
105	-3.3669
120	-3.0187
135	-2.4648
150	-1.7428
165	-0.90216
180	-1.3409E-8

```
Torque-angle characteristics for rotary machine
-------------------------------------------------
(assuming cosinusoidal inductance variation with angle)


Instantaneous stator current (A) ?10

Instantaneous rotor current (A) ?10

Stator self-inductance when stator
and rotor coils aligned (mH) ?50

Is rotor Salient or Round (S/R) ?S

Stator self-inductance when stator
and rotor coils orthogonal (mH) ?30

Rotor self-inductance when rotor
and stator coils aligned (mH) ?30

Is stator Salient or Round (S/R) ?R

Coefficient of coupling between coils when aligned ?.9


Angle (deg)         Torque (Nm)
-----------         -----------

    0               0
   15               -1.40216167
   30               -2.60886791
   45               -3.46475151
   60               -3.88471717
   75               -3.86691318
   90               -3.48568502
  105               -2.86691318
  120               -2.15266637
  135               -1.46475152
  150               -0.87681711
  165               -0.402161673
  180               -5.71515981E-9
```

Program notes
(1) The program assumes the rotor angle to be measured from the coils-aligned position, as indicated in Figure 6.8. With the variable names assigned, the various inductance variations are as indicated in Figure 6.13(a). Note that, with a round rotor, stator self inductance has the constant value L1, and that with a round stator, rotor self-inductance has the constant value L2. These values are input at lines 180 and 310 respectively. When the appropriate saliences are present, the minimum (coil orthogonal) stator and rotor self-inductancies, L0 and L3, are input at lines 260 and 390 respectively.
(2) Checks are made to ensure that a valid value of coupling coefficient (K1) is input (lines 420–450). The mutual inductance in the

coils-aligned position is then derived, using Equation (3.19) (line 460).

(3) A loop is set up, spanning lines 520–580, to calculate and display torque at 15° intervals from 0° to 180°. Referring to Figure 6.13(a), the interaction force component of the torque is calculated at line 540 from the third term of Equation (6.32). If the rotor is salient, an alignment component due to stator inductance variation follows from the first term. This is added at line 550. Similarly, for stator saliency if present, the rotor inductance variation alignment component is calculated from the second term and added at line 560.

(4) The two runs show the effect of adding rotor saliency to a round-stator machine. The results are displayed graphically in Figure 6.13(b). Note that the torque is negative. What is the significance of

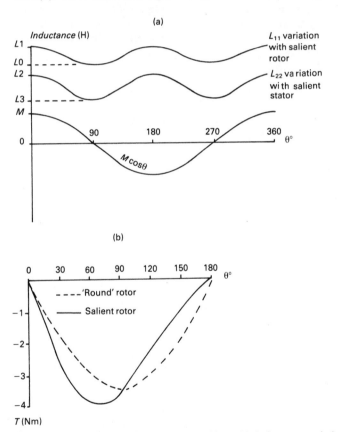

Figure 6.13 Performance of basic rotary machine   (a) inductance variations with angle   (b) torque-angle results for Example 6.3

this, and where will the rotor settle if it is free to rotate and is released
with the coils orthogonal? If steady values of positive current are
flowing in each coil, does rotor saliency increase or decrease the coil
misalignment when a small load torque is applied? (Study Figure
6.13(b).)

Example 6.4: ENERGY: ideal 'slow' operation of magnetically non-linear relay

Develop a general program to calculate the energy conversion characteristics of a d.c. relay. Magnetic leakage and fringing may be
neglected, but the effects of magnetic saturation should be included.
Constant m.m.f. (ideal 'slow') operation may be assumed. The circuit
material is to be mild steel, and input data required includes the dimensions of the circuit and airgap, and the applied m.m.f. Values of
flux density and force at the beginning and end of operation should
also be displayed.

```
10 REM      ENERGY - Energy balance in d.c. relay with
20 REM      ideal "slow" operation and magnetically
30 REM      non-linear iron, neglecting leakage and
40 REM      fringing. (based on rate of change of
45 REM      magnetic co-energy with distance at F const)
50 DATA 0,0, 0.2,100, 0.4,150, 0.6,200, 0.8,250, 1.0,400
60 DATA 1.1,500, 1.2,600, 1.3,800, 1.4,1300, 1.5,2000
70 DATA 1.55,2500, 1.6,3400, 1.65,4500, 1.7,6000
75 DATA 1.75,7500, 1.8,10000, 2.0,30000
80 DIM I1(17,1)
90 FOR J = 0 TO 17
100  FOR K = 0 TO 1
110   READ I1(J,K)
120  NEXT K
130 NEXT J
140 PI = 3.14159265
150 M0 = 4*PI*1E-7
160 DIM B3(1), F1(1), P3(1), W1(1), W2(1), X(1)
170 FOR J = 0 TO 1
180   W1(J) = 0
190  NEXT J
200 PRINT
210 PRINT
220 PRINT "Energy balance in d.c. relay with ";
225 PRINT "saturating iron"
230 PRINT "-----------------------------------";
235 PRINT "----------------"
240 PRINT "(ideal 'slow' operation assumed)"
250 PRINT
260 PRINT
270 PRINT "Total length of iron circuit (mm) ";
280 INPUT L1
290 L1 = L1*1E-3
300 PRINT
310 PRINT "Length of ONE airgap in OPEN position (mm) ";
```

```
320 INPUT X(0)
330 PRINT
340 PRINT "Length of ONE airgap in CLOSED position (mm) ";
350 INPUT X(1)
360 IF X(1)>X(0) THEN GOTO 930
370 FOR J = 0 TO 1
380    X(J) = 2*X(J)*1E-3
390    NEXT J
400 PRINT
410 PRINT "Circuit cross-sectional area (mm^2) ";
420 INPUT A
430 A = A*1E-6
440 PRINT
450 PRINT "Number of excitation turns ";
460 INPUT N
470 PRINT
480 PRINT "Excitation current (A) ";
490 INPUT I
500 F = N*I
510 V1 = M0*F/X(1)
520 V2 = V1*L1*I1(17,1)/F
530 IF I1(17,0) < V1 - V2 THEN GOTO 970
540 FOR J = 0 TO 1
550   FOR K = 1 TO 12
560     B1 = K*M0*F/(12*X(J))
570     H1 = K*F/(12*L1)
580     FOR L = 0 TO 17
590       IF I1(L,0) > B1 - (B1/H1)*I1(L,1) THEN GOTO 610
600       NEXT L
610     X1 = I1(L,1) - I1(L-1,1)
620     Y1 = I1(L,0) - I1(L-1,0)
630     D = Y1 + (B1/H1)*X1
640     B2 = B1/H1*(I1(L-1,0)*I1(L,1)-I1(L-1,1)*I1(L,0))/D
645     B2 = B2 + B1*Y1/D
650     P2 = B2*A
660     IF K=12 THEN W1(J) = W1(J) + P2
670     IF K=12 THEN GOTO 700
680     IF (-1)^K < 0 THEN W1(J) = W1(J) + 4*P2
690     IF (-1)^K > 0 THEN W1(J) = W1(J) + 2*P2
700     P3(J) = P2
710     B3(J) = B2
720     NEXT K
730   W1(J) = W1(J)*F/36
740   F1(J) = A*B3(J)^2/M0
750   W2(J) = F*P3(J) - W1(J)
760   NEXT J
770 PRINT
780 PRINT
790 PRINT "System co-energy (OPEN) = "; W1(0); " J";
795 PRINT "     (CLOSED) = "; W1(1); " J"
800 PRINT
810 PRINT "System energy (OPEN) = "; W2(0); " J";
815 PRINT "     (CLOSED) = "; W2(1); " J"
820 PRINT
830 PRINT "Flux density (OPEN) = "; B3(0); " T";
835 PRINT "     (CLOSED) = "; B3(1); " T"
840 PRINT
850 PRINT "Force (OPEN) = "; F1(0); " N";
855 PRINT "     (CLOSED) = "; F1(1); " N"
860 PRINT
```

```
870 PRINT "Electrical energy input = ";
875 PRINT F*(P3(1) - P3(0)); " J"
880 PRINT
890 PRINT "Work done at constant m.m.f = ";
895 PRINT W1(1) - W1(0); " Nm"
900 PRINT
910 PRINT "Change in stored field energy = ";
915 PRINT W2(1) - W2(0); " J"
920 STOP
930 PRINT
940 PRINT "Airgap in closed position must be shorter ";
945 PRINT "than in open"
950 PRINT "position. Please enter airgap lengths again"
960 GOTO 300
970 PRINT
980 PRINT "Load line does not intersect B-H ";
985 PRINT "characteristic when relay"
990 PRINT "closed. Reduce coil excitation (Amps x Turns)"
1000 GOTO 440
1010 END
```

```
Energy balance in d.c. relay with saturating iron
-------------------------------------------------
(ideal 'slow' operation assumed)


Total length of iron circuit (mm) ?250

Length of ONE airgap in OPEN position (mm) ?2

Length of ONE airgap in CLOSED position (mm) ?1

Circuit cross-sectional area (mm^2) ?100

Number of excitation turns ?1000

Excitation current (A) ?2


System co-energy (OPEN) = 6.1E-2 J   (CLOSED) = 0.119 J

System energy (OPEN) = 6.15E-2 J   (CLOSED) = 0.115 J

Flux density (OPEN) = 0.612 T   (CLOSED) = 1.17 T

Force (OPEN) = 29.8 N   (CLOSED) = 108 N

Electrical energy input = 0.111 J

Work done at constant m.m.f = 5.75E-2 Nm

Change in stored field energy = 5.35E-2 J

STOP at line 920
```

```
Energy balance in d.c. relay with saturating iron
------------------------------------------------------
(ideal 'slow' operation assumed)


Total length of iron circuit (mm) ?250

Length of ONE airgap in OPEN position (mm) ?2

Length of ONE airgap in CLOSED position (mm) ?1

Circuit cross-sectional area (mm^2) ?100

Number of excitation turns ?1000

Excitation current (A) ?5

System co-energy (OPEN) = 0.377 J   (CLOSED) = 0.59 J

System energy (OPEN) = 0.345 J   (CLOSED) = 0.297 J

Flux density (OPEN) = 1.44 T   (CLOSED) = 1.77 T

Force (OPEN) = 166 N   (CLOSED) = 250 N

Electrical energy input = 0.165 J

Work done at constant m.m.f = 0.213 Nm

Change in stored field energy = -4.8E-2 J

STOP at line 920
```

Program notes
(1) The program evaluates mechanical work done from Equation
(6.35) for the change in co-energy at constant m.m.f. This requires the
development of Φ–F characteristics for the relay in the open and
closed positions. The energy balance, flux densities and forces then
follow directly.
(2) **B–H** data for mild steel is held in two-dimensional array I1 in an
identical manner to that employed for Example 4.2 (ROTORB).
(3) Values of the various quantities required are stored in two-
dimensional arrays with subscripts 0 and 1 referring to the start
(open) and finish (closed) positions respectively. The arrays B3, F1,
P3, W1, W2, and X are assigned values of flux density, force, flux, co-
energy, energy, and airgap length.
(4) Checks are incorporated to ensure consistent airgap lengths (line
360), and that the load line generated for the closed relay (i.e. the
maximum flux density condition) intersects the **B–H** curve. This second check spans lines 500–530 and is similar to that used in Example
4.1 (CCORE).

(5) The main program loop evaluating the field co-energies for the open and closed positions covers lines 540–760. The circuit m.m.f. *F* (line 500) is subdivided into 12 equal parts. For each subdivision, the circuit flux density (B2) and flux (P2) is calculated using the linear interpolation technique employed in CCORE (lines 580–650). Simpson's rule is then used to evaluate Equation (6.34) numerically. (Simpson's rule has previously been employed in Example 2.4 (WAVES), and is described in detail in Mason (1983)). This covers lines 660–690 and line 730. The circuit flux and flux density in the open and closed positions is assigned to the arrays P3 and B3 at lines 700 and 710. The force (F1) follows from Equation (6.24), allowing for the two active airgap areas in the relay. The energy (W2) is finally derived at line 750.

(6) The electrical energy input is calculated and printed at line 870. Equation (6.8) is used which, although originally derived for the linear case, is clearly equally applicable when the iron saturates.

(7) The two runs for an identical relay indicate the effect of saturation on performance. In the first, with a maximum flux density of 1.17 T, saturation is negligible, and system energy and co-energy values are similar in the relay-open position. After the relay has closed, the two values have increased by comparable amounts. Additionally, the '50–50' rule is approximately obeyed, since the electrical energy input splits nearly equally between the magnetic field and the work output. In the second run, the current is 2.5 times higher, introducing significant saturation. The co-energy now exceeds the energy, and the difference increases from the open to the closed position. In fact, the system energy reduces during the movement, and so the work done exceeds the electrical energy input as additional energy is extracted from the magnetic field.

PROBLEMS

(6.1) Develop a program to analyse an ideal magnetically-linear attracted-armature relay, of the form described in Section 6.3. The program should calculate and display electrical energy input, mechanical work output, change in stored field energy, and average force. Input data should include the geometrical details of the relay, d.c. current and coil turns, open and closed airgap lengths, and whether 'very fast' or 'very slow' operation is required.

(6.2) A form of *moving-iron ammeter* may be produced, equivalent to a machine with the layout of Figure 6.8(c). The current to be measured passes through the 'round' stator coil, and the salient rotor, which in this case has no winding, carries a pointer to display the deflection. For zero current $\theta = 45°$ and rotation of the rotor is restrained by a spring with a linear torque-angle characteristic. The spring produces zero torque when $\theta = 45°$ and peak restoring torque when $\theta = 0°$. The self-inductance L_{11} of the stator coil is given by

$$L_{11} = 0.25 + 0.05 \cos(2\theta) \text{ H} \tag{6.48}$$

If the meter is to measure direct current up to a value of 10 A, write a program to calculate scale intervals for various choices of spring constant.

Are there any potential problems attendant with the use of this instrument for d.c? If the meter is subsequently used to measure a.c. values, will the reading indicate mean value or r.m.s. value of the alternating current?

(6.3) A linear actuator consists of two concentric coils, the inner (coil 1) fixed, and the outer (coil 2) moving, wound on a short cylindrical iron core. Self and mutual inductance values for the arrangement are as follows

$$L_{11} = 0.1 \tag{6.49}$$

$$L_{22} = 0.1e^{-20x} \tag{6.50}$$

$$M = 0.08e^{-10x} \tag{6.51}$$

where x varies between 0 and 0.1 m, and is the distance of travel of the moving coil. The coils carry direct currents between 0 and 5A in the same direction. Write a program to display interaction, alignment and total force acting on the moving coil over the allowable range of movement, for any valid pair of values of the two currents.

What would be the effect of changing the polarity of one of the currents?

(6.4) Program ENERGY of Example 6.4 calculates mechanical work using a co-energy approach. However, airgap flux density and then force may be readily calculated in the given arrangement for any airgap. Hence develop an alternative program to calculate the energy values for the same problem, in which mechanical work is calculated from the integral of force with distance. The input data section of the program should be unchanged, and the work integral should be evaluated numerically using Simpson's Rule. Compare results for various sets of input data with those given using program ENERGY.

(6.5) Write the complementary program to ENERGY to calculate the energy balance in the relay with ideal 'fast' operation. Mild steel with the same **B–H** characteristic should be assumed. Work done now follows from decrease in field energy at constant Φ. Electrical energy input is somewhat more difficult to calculate in this case.

Compare the mechanical work done in the ideal 'slow' and ideal 'fast' cases. Does saturation have a significant effect on the comparitive values obtained?

Index

ABS, 11, 93
Admittance, 51
  in parallel, 53
  matrix, 52, 69
Airgap, 73
Alcomax III, 86, 99–101
ALGOL, 1
Ampère circuital law, 76
Ampère-turn balance, 106
Argand diagram, 12–15
Array, 7–8
Assignment, 2–3
ATN, 11, 29–30
Attracted-armature relay, 139
  dynamic performance of, 151–157
  force developed by, 142–143
  magnetically non-linear, 166–170
  very fast operation of, 141
  very slow operation of, 141
  work done by, 140
Autotransformer, 120–121

Barium titanate, 24
BASIC
  advantages of, 1
  development of, 1
  minimal, 1
B–H curve, 75–76
$(BH)_{max}$ point, 101
Bilateral impedance, 44

Capacitance, 24
Capacitive reactance, 25
C-core, 73–74, 77, 87–91
Ceramic magnet, 86
Charge, 18, 22–24
Coefficient
  magnetic flux leakage, 74, 87–88, 91
  of coupling, 46
Co-energy, 149–150, 169–170

Coercive force, 86
Compiled language, 1
Complex number
  addition of, 14–15
  argument of, 13
  basic (Cartesian) form of, 12, 28
  conjugate of, 15, 56
  definition of, 12
  division of, 14–15
  exponential form of, 13
  imaginary part of, 12
  modulus of, 13
  multiplication of, 14–15
  polar form of, 13, 28
  real part of, 12
  subtraction of, 14–15
Complexor, 16–17
Conditional operators, 5
Conductance, 18
Conductivity, 18
Constant, 3
COS, 11

DATA, 3–4, 10
Data validation, 60
Debugging, 9–10
DEF FN, 8–9, 11
Dielectric, 22–23
  strength, 38
DIM, 7, 10
Dot notation, 47
Duality, 52
Dust core, 83

Eddy-current loss, 80–83
Electric
  charge, 18, 22–24
  generator, 138
  motor, 138
Electrodynamic wattmeter, 157–161

173